T0100474

Quasi-Dimensional Simulation of Spark Ignition Engines

Alejandro Medina · Pedro Luis Curto-Risso
Antonio Calvo Hernández
Lev Guzmán-Vargas · Fernando Angulo-Brown
Asok K. Sen

Quasi-Dimensional Simulation of Spark Ignition Engines

From Thermodynamic Optimization to Cyclic Variability

 Springer

Alejandro Medina
Antonio Calvo Hernández
Departamento de Física Aplicada
Universidad de Salamanca
Salamanca
Spain

Pedro Luis Curto-Risso
Instituto de Ingeniería Mecánica y
    Producción Industrial
Universidad de la República
Montevideo
Uruguay

Lev Guzmán-Vargas
Instituto Politécnico Nacional Unidad
    Profesional Interdisciplinaria en
    Ingeniería y Tecnologías
México D.F.
Mexico

Fernando Angulo-Brown
Instituto Politécnico Nacional de Física y
    Matemáticas
México D.F.
Mexico

Asok K. Sen
Department of Mathematical Sciences
Indiana University
Indianapolis
USA

ISBN 978-1-4471-5288-0   ISBN 978-1-4471-5289-7   (eBook)
DOI 10.1007/978-1-4471-5289-7
Springer London Heidelberg New York Dordrecht

Library of Congress Control Number: 2013940296

© Springer-Verlag London 2014
This work is subject to copyright. All rights are reserved by the Publisher, whether the whole or part of the material is concerned, specifically the rights of translation, reprinting, reuse of illustrations, recitation, broadcasting, reproduction on microfilms or in any other physical way, and transmission or information storage and retrieval, electronic adaptation, computer software, or by similar or dissimilar methodology now known or hereafter developed. Exempted from this legal reservation are brief excerpts in connection with reviews or scholarly analysis or material supplied specifically for the purpose of being entered and executed on a computer system, for exclusive use by the purchaser of the work. Duplication of this publication or parts thereof is permitted only under the provisions of the Copyright Law of the Publisher's location, in its current version, and permission for use must always be obtained from Springer. Permissions for use may be obtained through RightsLink at the Copyright Clearance Center. Violations are liable to prosecution under the respective Copyright Law. The use of general descriptive names, registered names, trademarks, service marks, etc. in this publication does not imply, even in the absence of a specific statement, that such names are exempt from the relevant protective laws and regulations and therefore free for general use.
While the advice and information in this book are believed to be true and accurate at the date of publication, neither the authors nor the editors nor the publisher can accept any legal responsibility for any errors or omissions that may be made. The publisher makes no warranty, express or implied, with respect to the material contained herein.

Printed on acid-free paper

Springer is part of Springer Science+Business Media (www.springer.com)

*The work devoted to this book is dedicated
with the warmest affection to my family,
Begoña, Miguel, and Macarena*

A. Medina

*I dedicate this book to my family, especially
Christine for all her support in all
my projects and Matías just for being
a definite part of our life*

P. L. Curto-Risso

In memory of my parents, Rosa and Antonio

A. Calvo-Hernández

Foreword

The road to this book began around 7 years ago after reading a paper by A. Fischer and K.-H. Hoffmann, *Can a Quantitative Simulation of an Otto Engine be Accurately Rendered by a Simple Novikov Model with Heat Leak?*, J. Non-Equilib. Thermodyn., 2004, 29, 9–28. We were by then interested in models from the Classical Equilibrium Thermodynamics that starting from an ideal Otto cycle were capable to lead to efficiencies closer to that of real spark ignition engines. This paper gave us a lot of ideas in order to use different kinds of models for these engines. We tried to develop realistic models beyond the Otto reversible cycle in which the main physical ingredients of real engines could be thoroughly analyzed. This aim leads us to quasi-dimensional simulations. We found a lot of research papers and works, and some sections in internal combustion engines textbooks, but we thought that there was not a monographic book explaining in detail which should be the main structure of a quasi-dimensional simulation model and how to arrange the basic submodels required to obtain numerical results. We had to deal with several practical points and also to elect, for each submodel, which one could be the most adequate for our purposes. This made the development of simulations not so straightforward as we thought at the earlier stages and reinforced our feeling about the eventual usefulness of a monograph on this kind of simulation models.

This text is intended for students and researchers, that without an extensive knowledge on spark ignition engines, are interested on understanding the physical, chemical, and engineering basis of these engines. And going further those are interested to develop their own simulation scheme or to have a deep understanding of which is behind commercial simulation software. We have tried to write the book for readers with merely a basic background on thermodynamics, mechanics, chemistry, and numerical methods. This text attempts to be an introductory guide, in such a way that the interested reader could deep in details about submodels, other models not directly considered, about validations, or about the kind of information that could eventually be extracted from computations by using the cited references. We have tried to be careful in presenting broad and updated bibliography resources to give the reader a starting point for going beyond depending on his own interests.

Internal combustion engines, and specifically spark ignition engines, are one of the most important devices designed and developed by scientists and engineers for our current lifestyle. Associated to fossil fuels resources limitations and to the increasing environment concerns, these engines have shown during the last years an important potential for improvements, linked to their design and optimization. Different kind of simulations play a basic role in such work. Quasi-dimensional simulations, with some limitations (for instance when compared with fluid dynamic models), are attractive because of their simplicity and because they do not require extensive computer facilities or much computing time. And moreover, they allow to identify, in quite a direct way, the influence of the basic physical and chemical hypothesis on engine functioning. Probably because of the previous background and interests of some of the authors, we paid special attention to thermodynamic considerations, that are in the basis of most energy conversions. The optimal design of such processes, as for example the chemical to mechanical energy conversion occurring in internal combustion engines, is in the basis of any energy-saving technology.

Salamanca, April 2013   Alejandro Medina

Acknowledgments

More than once we, the authors of this book, during the writing process joked by commenting that any book on internal combustion engines would be only a kind of (bad) summary of the well-known reference book by J.B. Heywood, *Internal Combustion Engine Fundamentals*. As most of the researchers on this field, we are indebted to him because of his outstanding contributions and writings.

A. Medina and A. Calvo Hernández acknowledge Ministerio de Ciencia e Innovación, Junta de Castilla y León, and Universidad de Salamanca (Spain) from their financial support during the last years. We also would like to express our gratitude to all the members of the Research Group on Thermodynamics and Statistical Mechanics, Universidad de Salamanca, for their continuous support and encouragement.

P. L. Curto-Risso would like to thank Agencia Nacional de Investigación e Innovación (Uruguay) for the SNI program that supports his research. Also thanks to his colleagues in the Department of Applied Thermodynamics from Universidad de la República and his students, Bruno, Diego, and Daniel, who have contributed to deeply analyze different heat transfer models.

Lev Guzmán-Vargas and F. Angulo-Brown thank the financial support from COFAA-IPN, EDI-IPN and Consejo Nacional de Ciencia y Tecnología (Project 49128–26020), México.

Contents

Chapter 1
Introduction

1.1 Modeling Spark Ignition Engines

In addition to experimental techniques, modeling is nowadays a basic procedure for
the design, development, and optimization of internal combustion engines. Besides,
modeling is a method to improve our knowledge of the physics and chemistry underlying their operation. A combination of simplifying hypotheses, basic physical and
chemical laws expressed in the form of mathematical equations, and the numerical
computational capabilities of computer facilities, enable us to propose realistic models, validate them by comparing with real engine outputs, and estimate the sensitivity
of the engine variables to several parameters. As stated by Heywood [1], there are at
least four main objectives in modeling engines:

1. To improve our knowledge about the main physical and chemical ingredients
   which govern the functioning of engines.
2. To analyze the key parameters of engine design and operation. This could be specially valuable in order to provide guidelines for more rational and less expensive
   experimental development efforts.
3. To predict the sensitivity of the engine to several control variables that could lead
   to optimize engine design, control, and operation.
4. To predict the effect of design innovations before their experimental development
   and testing in bench engines.

It is difficult to assert which are the best models. Different models have different accuracy depending on the objective to analyze. Models developed for internal
combustion engines are usually divided into two main categories, depending on the
physical basis that is used to state the basic equations. These are *thermodynamic* and
fluid dynamic models. Within thermodynamic models, the terminology of *single-
zone* (or *zero-dimensional*) and *multi-zone models* is often used. Single-zone models
are based on the conservation of mass and energy through the first law of thermodynamics and have no spatial resolution, because all the thermodynamic properties
are considered as uniform, and one single control volume is taken. In multi-zone

A. Medina et al., *Quasi-Dimensional Simulation of Spark Ignition Engines*,
DOI: 10.1007/978-1-4471-5289-7_1, © Springer-Verlag London 2014

models, the combustion chamber is divided into several zones, with each zone possessing uniform thermodynamic properties. The first law of thermodynamics is applied to each of these control volumes with appropriate boundary conditions. The mathematical equations form a set of ordinary differential equations whose independent variable is the time or the crank angle. *Quasi-dimensional models* are a particular type of these models. Fluid dynamic models, also known as Computational Fluid Dynamics (CFD) models, are inherently multi-dimensional models which are based on the conservation of mass, chemical species, momentum, and energy at any location within the engine cylinder. Mathematically, the starting equations of this type of models are the Navier–Stokes equations, which constitute a system of partial differential equations. Both time and spatial coordinates are considered as independent variables, so it is possible to get a full spatial resolution of the properties of the gas inside the cylinder. In the following subsections, we describe the basic features of these models as well as their range of applicability.

1.1.1 Single-Zone (or Zero-Dimensional) Models

These thermodynamic models are based on the thermodynamic analysis of the cylinder contents during the engine operating cycle. In these models, a general expression for the first law of thermodynamics is applied to an open system formed by fuel, air, and residuals sequentially for each stroke. In some strokes (compression and expansion), the thermodynamic system can be considered as closed, and in others (intake and exhaust) the mass flow through the valves should be taken into account. Similarly, the chemical composition of the gas mixture is not identical in all the strokes, since it may contain a fuel-air-residuals mixture (before combustion) or only residuals[1] (after combustion).

In these models, the only independent variable is time. Spatial coordinates do not appear in the evolution of the system, and thus these models have no spatial resolution. So all the thermodynamic properties are homogeneous in the considered control volume. This is the reason these models are also called *zero-dimensional* models. The governing equations are ordinary differential equations with some initial conditions for each stage [1–3]. These models incorporate empirical laws for combustion through the mass burning rate. It is usual to take either of two analytical functions depending on some phenomenological parameters: a Wiebe function or a cosine law [4–6]. The details of the geometry of the combustion chamber and the geometry of the flame front do not play an explicit role in the evolution of the engine variables.[2] Within these models, the knowledge of the empirical burning rate has to be taken as an input data, calculated from prior experiments. Extrapolation for other conditions can be problematic. When the burning rate is known under the given

---

[1] Among residuals could exist a mass of unburned fuel.

[2] Some of the parameters appearing in the empirical burning rates may depend on engine geometry and other kind of data as rotational speed, fuel type, or fuel-air ratio.

conditions, these models are computationally easy to apply and allow us to obtain accurate results for global parameters such as the engine efficiency or power output.

With respect to the prediction of emissions, these models can estimate hydrocarbons (HC) or CO emissions due to partial combustion but not those arising from temperature variations inside the combustion chamber. Apart from these limitations, single-zone models can provide an estimation of NO_x emissions, because these are essentially associated with the temperature of the hottest regions of the combustion chamber, which is close to the mean gas temperature of the mixture [7]. Furthermore, these models can be used to check which mechanisms are the most important in the fuel-air oxidation reactions. This may be of interest, for instance, in the case of alternative fuels [8–12].

1.1.2 Multi-Zone Models

Some of the shortcomings mentioned above for single-zone models can be overcome with the use of multi-zone models. In these models, the combustion chamber is divided into a certain number of zones, each of them having uniform thermodynamic properties. As we shall discuss below, most usual multi-zone models consider two zones, but any number of zones may be used. The optimum number of zones is determined by means of a sensitivity analysis, i.e., by looking for the minimum number of zones that guarantees the convergence of the simulation results [7]. The first law of thermodynamics is applied to each zone and it is assumed that adjacent zones exchange heat, work, and mass with each other. The temperature, composition, and any other thermodynamic property of each zone is considered uniform within the boundaries of the zone, but they can be different from the thermodynamic properties of other zones. This approach permits to analyze the effects of temperature and species stratification, which leads to a better estimation of combustion duration, pressure rise during combustion, and exhaust emissions.

The general formulation for a multi-zone model can, in principle, handle any specified number of zones, as long as all the terms expressing exchange of properties among the zones can be reasonably modeled. With many zones these quantities could be difficult to estimate if the velocity field is not found with spatial details. So, in practice, sometimes only a few zones for the bulk gases are considered. The additional zones are applied to the wall boundary layers and the crevice volumes. These zones are important for understanding the emissions and heat losses in the engine.

A simpler, but widely applied methodology is a *two-zone* model analysis of the engine. These models are also known as *quasi-dimensional* models. Quasi-dimensional models were developed to bridge the gap between zero- and multi-dimensional models [1, 13–15]. They introduce a spatial dependence in processes such as combustion and heat transfer. Instead of deriving this spatial dependence from the solution of a full set of partial differential equations as multi-dimensional models do, there are some phenomenological foundations that are often used.

These models are useful because, with modest computing times, they can relate the model outputs to the combustion chamber geometry.

The two zones considered in quasi-dimensional simulations are only effective during combustion, where they are selected as the unburned and burned gas volumes. The separation between the zones is the propagating flame. For spark ignition engines, experimental photographic recordings reveal that the two zones are reasonably well formed. Experiments also show that the flame front is nearly spherical and with a relatively thin structure [6, 16–18]. In engines with *swirl* or complex geometry the center of the spherical flame front may not be at the spark plug, but at a point that changes with time. The flame propagation is the rate at which the unburned mixture is converted into the burned gases. Except in the case radiation heat transfer is considered, it is generally assumed that there is no heat transfer between the zones.

There exist several mathematical models of turbulent flame propagation leading to a set of differential equations for the evolution of the masses of the burned and unburned gases [13, 19–21]. During combustion, these equations are solved simultaneously with the thermodynamic ones for the temperatures and pressure in the cylinder. The combustion chamber geometry plays an important role in the evolution of the system during combustion. Throughout all the other phases of the cycle there is only one zone present in the chamber, although its chemical composition, either burned or unburned gases, depends on the stroke. For solving the thermodynamic equations in all the strokes several submodels are required, for instance, for heat transfer, frictions, mass flows through the valves, chemical reactions, and so on. Some of these submodels arise from basic physical or chemical laws and others are empirical in nature.

In quasi-dimensional models it is possible to analyze different engine geometries, rotational speeds, fuels and emissions, and many other design or operating parameters of the engine. More evolved or complex submodels can be incorporated into the simulations in order to analyze specific effects. As in zero-dimensional models, the only independent variable is time, so solving partial differential equations systems is not required. This leads to computationally inexpensive models, capable of performing precise sensitivity studies of the engine, and also to analyze cycle-to-cycle variations. Conceptually, these models are not complicated, so they allow us to check and identify the importance of several physical and chemical mechanisms in the evolution of the engine parameters. Among their limitations, we should mention that all those quantities which are affected by spatially resolved properties are difficult to estimate.

The primary objective of this monograph is to describe the main ingredients that a quasi-dimensional model should incorporate and to show the capabilities of these models for engine optimization and for the prediction and analysis of the cycle-to-cycle variations which are observed in experimental studies.

1.1.3 Multi-Dimensional Models

Although multi-zone models are associated with some spatial considerations, they usually do not allow high local resolution. Particular problems as the analysis of flows through valves, fuel injection, bulk in-cylinder flow, and turbulent development do require a special treatment with a detailed spatial resolution [3, 7, 22]. Multi-dimensional or Computational Fluid Dynamics (CFD) based models describe in detail the charge motion inside the combustion chamber and the turbulence effects, and provide a high spatial resolution within the combustion chamber.

Since the 1960s, the aerospace industry has integrated CFD techniques into the design, research and development of aircraft, and jet engines. More recently, these methods have been applied to the design of internal combustion engines and gas turbines [23]. Their development during the past few years has been linked to the evolution of modern computer capabilities. Simultaneously, a great deal of effort has been devoted to develop more efficient numerical algorithms and to provide easy interfaces for the input and output [24]. Despite such developments, the computation requirements of CFD models are much larger than those of the one- or multi-zone techniques we described before.

CFD computational codes are structured around the numerical algorithms that can tackle fluid flow problems. They solve the partial differential equations for conservation of mass, momentum (Navier-Stokes equations), energy, and species concentrations for a discretized representation of the combustion chamber and the inlet and exhaust ducts. The region of interest is divided into a number of small zones or cells that form a grid or mesh. Only under special circumstances, a two-dimensional axisymmetric representation is possible: usually a full 3D representation is required. The discretization procedure is a basic step in these kinds of models, as well as the definition of the geometry of the region of interest, the definition of the fluid properties, and the specification of the appropriate boundary conditions. The accuracy of the solution and its computational cost (necessary computer hardware and computation time) are dependent on the fineness of the grid.

There are four different types of numerical solution techniques: finite difference, finite element, finite volume, and spectral methods (see [24] for details). Within any of these techniques the solver performs the following steps: approximation of the unknown flow variables by means of simple functions, discretization by substitution of the approximations into the governing flow equations and subsequent mathematical manipulations, and solution of the corresponding algebraic equations. The main differences among the different solution techniques are associated with the way in which the flow variables are approximated and with the discretization processes.

1.2 Thermodynamic Optimization

The energy conversion processes in real energy converters are always linked to unavoidable losses occurring as entropy is produced. These losses limit the efficiency of such processes. However, in real heat engines typically a few sources of irreversibility overwhelmingly dominate losses [25, 26]. About 60 years ago, the effect of finite-rate heat transfer was related to the efficiency of heat engines [27, 28]. During the oil crisis of 1970s, the framework of finite-time thermodynamics evolved as a theoretical scheme to determine performance bound for thermodynamic processes proceeding in a finite time or with finite rates [29–32]. One particular effective (and also criticized [33–35]) development were the endoreversible models. Curzon and Ahlborn [36] showed that a more realistic bound (compared with Carnot's limit) for the efficiency of heat engines operating at maximum power can be calculated when accounting for the irreversibility associated with a finite-rate heat transfer. The heat engine is modeled as internally irreversible (endoreversible) with all irreversibilities coming from the heat exchange with the external reservoirs [37]. From then, *finite-time thermodynamics* and, in general, *thermodynamic optimization* have been applied to several energy converters, from the macroscopic usual engines and refrigerators, to systems with mesoscopic or microscopic scales [38–42].

Finite-time thermodynamics (FTT) could be considered as an extension of Classical Equilibrium Thermodynamics (CET). It provides insights into realistic heat devices, working under isothermal and nonisothermal conditions, which are constrained to evolve in a finite time and to have a finite size. The focus is then placed on the analysis of the unavoidable irreversibility of the real processes. FTT basically proposes a thermodynamic model incorporating the main irreversibilities coming from the system itself and from its coupling with the external world. The system is considered as a whole (only one control volume), so to some extent this kind of models is a particular type of the zero-dimensional models discussed before.

Within this thermodynamic model, an optimization of the system is performed. In particular, a figure of merit or an objective function is proposed, and optimization of the entropy generation and exergy are carried out. References [25, 32, 43–47] are well-known reviews where detailed information is given on models and sources of irreversibility such as heat and mass transfer, fluid flow, turbulence, and combustion. More recent reviews include updated bibliography on heat engines and refrigerators, distillation, chemical reaction, optimal control and stability, and various other aspects.

The value of these simple models lies in their low computational cost and the fact that they predict energetic properties with a reasonable accuracy, providing deep physical insights on the constraints associated with any energy converter. In regards to internal combustion engines, most of the models consider as a starting point an air standard Otto, Diesel, dual-like cycles, etc., with piston-cylinder frictional losses, heat losses through the cylinder wall, and constant heat capacities. These minimum ingredients are sufficient to obtain the well-known loop-shaped power-efficiency curves which reflect the performance characteristics of a real heat engine. Usually, the emphasis is put on the maximum efficiency and/or maximum power regimes, although different figures of merit based on the optimization of a certain compromise have been analyzed [48, 49]: the ecological criterion (compromise between maximum power output and minimum entropy generation), power density (between power output and size), and maximum efficient power (between power and efficiency). In addition, universal features of heat engines operation are cast in the form of power versus efficiency plots, which allow us to identify the optimal operating regimes [25].

On the above basic scheme, some recent and more elaborated works on spark-ignition engines include the following improvements [50–55]: (a) analysis of the influence of nonconstant heat capacities of the working fluid; (b) consideration of internal irreversibilities associated with the characteristics of the chemical reactions; (c) effects of variable speeds and of the fuel–air equivalence ratio; and (d) the incorporation of cycle-to-cycle variability. Fischer and Hoffmann [26] explicitly compared zero-dimensional simulation results with a simple FTT model for a Carnot-like engine [37] and demonstrated that the theoretical model can reproduce the complex engine behavior by an appropriate choice of model parameters.

A second kind of analysis for internal combustion engines in FTT involves the optimization of the piston motion. These studies include the effects of finite piston acceleration from a generic model of expansion of a heated fluid inside a cylinder, and apply appropriate optimal control theories. In this type of models (see Refs. [56, 57] and references therein), the optimization procedure determines the optimal trajectory on each stroke as a function of time and then to optimize the total cycle time by maximizing a figure of merit. Special attention is paid on the mechanical friction (internal or external), on the heat transfer law (mainly newtonian, inverse, radiative, and/or different combinations). Details on the chemical reactions taking place in the combustion chamber can also be incorporated. Commonly used figures of merit are the maximum work output (for a fixed total cycle time), the maximum efficiency, and the minimum entropy generation.

In this book, we shall see how the utilization of finite-time thermodynamics techniques together with the quasi-dimensional simulation models allow us to obtain valuable insights into the operation of spark ignition engines. We will also propose optimization procedures to find improved design and operational parameters for the engines.

1.3 Cycle-to-Cycle Fluctuations

It is well-known from experimental studies that the pressure inside the cylinder of an internal combustion engine shows substantial variations among successive cycles. For a multi-cylinder engine, these variations occur in each cylinder and also from one cylinder to another. After extensive research on this subject, it has been recognized that the cycle-to-cycle fluctuations arise from the motion of the mixture before and during combustion, variations in the amounts of fuel, air, and residual gases inside the combustion chamber, and changes in the composition and homogeneity of the gas mixture in each cycle.

Generally, cyclic dispersion increases with any factor which slows down the combustion process, such as, lean mixtures, larger proportion of exhaust gas residuals, and low load operation. Nowadays, the primary objectives of vehicle manufacturers are the reduction of fuel consumption as well as reduction of exhaust emissions. Among the several strategies that are being used to accomplish these goals are fast combustion, lean burn, variable valve timing, direct injection, and exhaust gas recirculation. Several of these strategies may, however, result in increased cyclic variability. Thus, it is important to develop a good understanding of the physical and chemical basis of cyclic dispersion, with the final aim to diagnose and control it.

High levels of variability deteriorate engine driveability due to the large fluctuations in the torque and work output. For good driveability, the *coefficient of variation*[3] of the *indicated mean effective pressure* (IMEP) should not exceed 5–10 %. Several authors [58–62] claim that if cyclic dispersion is reduced, the torque and power output could increase a significant amount. In terms of fuel consumption, if all cycles burn at the optimum rate an appreciable improvement in fuel economy could be reached. At the other extreme, as Stone [59, 60] has argued, probably the total elimination of cyclic variability may not be desirable, because it would be more difficult to control *knock* under this condition.

Optimum spark timing is usually set for an average standard cycle, which means that for cycles with a combustion event faster than average, spark timing will be too advanced and conversely for slower cycles. Both circumstances lead to power and efficiency losses. Fast cycles are most likely to knock, so they impose constraints on the octane requirements of the fuel and limit the compression ratio. Slow cycles are susceptible to incomplete combustion, so they impose restrictions on the lean operating limit of the engine (or the level of exhaust gas recirculation). As a consequence, the cycle-to-cycle variability determines the range of spark advance angles and also the fuel-air ratios, which can be used.

The cycle-to-cycle variations can be analyzed in terms of several parameters such as in-cylinder pressure and heat release. Heywood [63] and Ozdor [61] have classified these parameters into four main groups.

1. Parameters related to pressure: the maximum cylinder pressure and the corresponding crank angle, the maximum rate of the pressure rise and the crank angle at which this occurs, and the indicated mean effective pressure.
2. Parameters related to the burn rate: the net heat release, the maximum mass burning rate, and the flame development or the rapid burning angle interval.
3. Flame front parameters: flame radius, flame front area, burned volume at given times, or flame arrival time at given locations.
4. Exhaust gas-related parameters: concentration of different components in the exhaust gases.

---

[3] The coefficient of variation is a classical statistical parameter that quantifies the dispersion of a noisy measure. It is defined as $COV = \sigma/\mu$ where μ is the mean value of the signal and σ its standard deviation.

Modern data acquisition systems can be used in the engine laboratory to monitor and measure these parameters very accurately.

1.3.1 Physical Origin of Cycle-to-Cycle Variations

The terms *cyclic variability* or *cycle-to-cycle variations* usually cover several types of fluctuations associated with their physical origin [61, 63, 64]:

(i) The variations in the motion of the gaseous mixture in the cylinder during combustion.
(ii) The variations of the amounts of fuel, air, and residual or recirculated exhaust gases in the cylinder.
(iii) The variations of the mixture composition and homogeneity, especially in the vicinity of the spark plug.
(iv) Fluctuations of some features of the spark discharge, like breakdown energy, and the initial flame kernel position.

The relative importance of these factors depends on the engine design and operating conditions. Overall, the early flame development is supposed to have substantial influence on the combustion process. The initial characteristics of the flame front will depend on the local fuel–air ratio, the exhaust gas residuals in the vicinity of the spark plug at the time of ignition, and the mixture motion in this region [61]. Next we briefly discuss the items listed above.

(i) Two main factors influence the variations of the gas motion (velocity field) within the cylinder throughout a cycle and from one cycle to the next: the mean flow velocity, and the turbulent velocity fluctuations [63, 65–68]. These fluctuations convect the flame at its initial stage to different directions, changing the flame development area, and the heat transfer with time. If the flame kernel is moved toward the thermal-boundary layer that covers the inner walls of the combustion chamber, it will burn slowly. Besides, a larger contact with the walls will reduce the flame front area. Nevertheless, if the initial flame kernel is moved away from the walls, it will burn more quickly. De Soete [69] and Lyon [58] have done detailed parametric studies about the consequences of the initial combustion stage on the whole combustion process. Lord et al. [70] have analyzed the effects of charge motion on the early flame kernel development. Winsor and Patterson [71] suggested that improving mixing turbulence can contribute to decrease the cyclic fluctuation rate. Several efforts have been made to analyze the consequence of *swirl* effects on cyclic fluctuations [72–75].
(ii) From one cycle to the next there may be variations in the amount of fuel, air, and residual gases in the cylinder [76–78]. Such variations may occur especially at low loads and for lean mixtures, where combustion variability is much higher and *partial burning* or even *misfires* can happen. In partial burning, the combustion process is so slow that it is not completed before the exhaust valve opens,

so a fraction of the injected hydrocarbons remain in the combustion chamber (together with reaction products) until the next cycle. For very dilute mixtures there can be a fraction of cycles that fail to ignite. Both situations result in a decrease in efficiency, higher hydrocarbons emissions, and torque and speed variations. The operating point at which engine evolution becomes rough and unstable and hydrocarbons emissions rapidly increase is usually called *stable operating limit* [63]. At higher loads and/or richer mixtures the combustion process is more repeatable and it is always completed early with respect to the power stroke and the variations in the amount of exhaust gases or unburned hydrocarbons are not significant.

(iii) Additionally, nonuniformities in the composition of the mixture before combustion are observed. The composition of the mixture changes, affecting the development of combustion especially in the region close to the spark plug. Effective fuel–air ratio in the vicinity of the spark plug may show a dispersion that affects all the subsequent processes [63]. The way the exhaust gases are mixed with the fresh mixture is a key point. Different *exhaust gas recirculation* (EGR) implementations, either external or internal, and *stratified charge injection* techniques would lead to different composition and homogeneity of the mixture [79–84]. Lean burn combustion techniques as well as EGR are focused on extending the maximum burning lean limit in order to reduce NO_x emissions to satisfy the emission restrictions [85, 86].

(iv) Other factors that influence cycle-to-cycle variations are the design of the ignition system, the number of spark plugs, and their orientation and location. In spark ignition engines, the electrical discharge between the electrodes of the spark plug starts combustion. The kernel created by the spark develops into a self-sustaining and propagating flame front (considered as a thin reaction region where the exothermic chemical reaction takes place).

The ignition system generates the required voltage (around 10^4 kV) across the electrodes to ignite the combustible mixture. This should be done under all operating conditions.[4] Usually spark timing, close to the end of the compression stroke, is set to give *maximum brake torque* (MBT timing), although concerns about emissions and knock control should be taken into account. The spark is usually provided by a battery and a coil, but for some applications a magneto is used [63, 87]. Among the several phases of the spark discharge (pre-breakdown and breakdown phases, arc phase and glow discharge phase), the breakdown phase[5] have the greatest effect on the ignition process. The size of the activated volume at a particular time after ignition, the temperature difference across the

---

[4] Although in terms of electrical energy about 0.2 mJ should be enough to ignite a quiescent stoichiometric fuel-air mixture under normal conditions, in real conditions, leaner or richer mixtures do require 30–50 mJ from the spark plug.

[5] Before the discharge occurs, the mixture in the cylinder behaves as a good insulator. At the spark pulse, the potential difference between electrodes rapidly increases and causes electrons to accelerate toward the anode (*pre-breakdown phase*). After a sufficient amount of electrons are released, the current through the gap quickly increases (up to around 100 A). This phase is known as *breakdown phase*.

kernel interphase, and the velocity of the interphase are considerably increased by increasing the breakdown phase energy. In the case of lean mixtures, the chemical energy density of the mixture is low and also the flame temperature. Consequently, the flame speed decreases and cycle-to-cycle variations tend to increase.

Methods to improve ignition have evolved from the traditional battery-coil systems to much more sophisticated electronic devices. For instance, laser ignition systems have been developed in order to ignite extremely lean fuel–air mixtures, which require high ignition energy [86]. These devices extend the engine stability limit and reduce the cyclic variations near the stability limit. A summary of these devices can be found in [63, 88–91].

1.3.2 Theoretical and Simulation Models

Besides experimental studies,[6] there have been considerable effort to develop simplified models which can be implemented in computer simulations in order to analyze the main physical ingredients in the nature of the cyclic variations.

Daw et al. [93, 94] developed a low-dimensional physically oriented model of cyclic variability in which the residual gas from each cycle alters the in-cylinder fuel-ratio through a pseudo-random component following a gaussian distribution. This can also be seen as a cycle-by-cycle variation of the *combustion efficiency*. Good quantitative agreement with experimental time-series measurements was found. Within the framework of FTT, Angulo-Brown et al. [55] presented a thermodynamic model, based on the Otto standard cycle but including explicit chemical reactions and a fluctuating heat release in the combustion process. Lean and rich mixtures were analyzed by considering different stochastic components in the heat release. Good qualitative agreement in several parameters was achieved when compared with experimental observations.

Among the simulation approaches, multi-dimensional models require large storage capacity and large computer times, particularly in cyclic variability computations [3, 95, 96]. Galloni [97] has analyzed the influence of the mean flow in the spark region by using CFD simulation codes. His results were obtained for 700 engine cycles. *Large-eddy simulations* (LES) have also been applied together with experimental results in order to acquire a better understanding of cyclic variability [98–100].

The drawbacks of CFD simulations in regards to computing time are avoided by quasi-dimensional models [16, 101]. It will be shown in Chap. 5, that quasi-dimensional models have the capability to reproduce, not only qualitatively but also quantitatively, the main characteristics of cyclic variations in spark ignition engines [64, 102–106]. All these models have the common feature of incorporating

---

[6] Several reviews about experimental analysis on cyclic variability in spark ignition engines can be found in the literature [61, 92]

different fluctuating components in any of the parameters characterizing the turbulence during combustion. The incorporation of such *pseudo-random* components at the early stage of combustion allows us to reproduce most features of the nonlinear evolution of the variables, such as pressures at different crank angles during combustion, net heat release, power output, and many others. An advantage of this kind of analysis is that the computer time required to obtain long time series is not large, and so the temporal organization of long time series (looking for memory effects, characteristic frequencies, long-range correlations, and so on) is practicable. It is possible to perform analysis over several thousands of engine cycles. Moreover, these models allow us to discern the importance of the basic sources of cyclic variations, since the main stages of the cyclic evolution of the engine are realistically implemented.

1.3.3 Non-Linear Dynamics Analyses

In the past few decades, there has been a growing interest in the study of systems which exhibit irregular time series. The presence of complex fluctuations is a common characteristic of systems frequently encountered in different areas such as biology, economy, physics, sociology, and engineering [107–111]. The identification of nonlinear characteristics in these systems is an important task to understand the behavior of the system under different operational conditions. The use of tools from nonlinear dynamics has become important for the analysis of irregular noisy signals, extracting new information which cannot be obtained by means of traditional methods. In many cases, the output reflects the participation of several mechanisms acting on different scales, leading to *fractal* properties. The fractal organization indicates that common features are present when spatial or temporal scales are varied. For instance, the presence of memory is frequently expressed by a power-law relationship between the correlation function and the lag, and the exponent characterizes the underlying dynamics. In other cases, the *correlation function* is related to the lag according to an exponentially decaying function with a characteristic scale in the memory. According to studies of correlation in time series, an integrated approach is more effective when the presence of memory is going to be evaluated.

In regard to the nonlinear dynamics involved in cycle-to-cycle variations in spark ignition engines, we mention the following experimental and theoretical works. Letelier et al. [112] reported experimental results for the variation of the in-cylinder pressure versus crank angle for a four-cylinder spark ignition engine. By reconstructing the *phase space* and making the *Poincaré sections*, it was concluded that cycle-to-cycle fluctuations are not governed by a *chaotic process* but rather by a superimposition of a *nonlinear deterministic dynamics* with a *stochastic component*. Daw et al. [93, 94] analyzed the results of a mathematical model for cyclic variations by means of *bifurcation plots* and *return maps* for different parameters of the model. Also, an analysis of the *temporal irreversibility* in time-series measurements was presented by using a symbolic approach to characterize the noisy dynamics [113, 114]. Scholl and Russ [115] also analyzed the *first-return maps* obtained from experimental measurements of the *gross mean effective pressure* at several fuel–air ratios.

Sen and coworkers [80, 116–121] have reported extensive work relative to cycle-to-cycle variations of peak pressure, peak pressure location, and heat release. Conventional *histograms* and *statistical moments* as well as *wavelet* decompositions were used to analyze the corresponding time series. The observed qualitative changes in combustion were analyzed by Litak et al. [122, 123] using different statistical methods such as return maps, *recurrence plots*, *coarse-grained entropy* [124, 125], *multi-fractal techniques* [116, 118], and *multi-scale entropy* (MSE) [126]. Curto-Risso et al. [104, 105] have applied monofractal and multifractal methods to characterize the fluctuations for several fuel–air ratios, from lean mixtures to stoichiometric situations. Time series were obtained by means of quasi-dimensional computer simulations incorporating a turbulent model for combustion. For intermediate mixtures a complex dynamics was observed, characterized by a crossover in the scaling exponents and a broad multifractal spectrum. Several works have been devoted to analyze the influence of different fuels (including alternative fuels) on variability [119, 121, 127–130] and to analyze the transition from spark ignition to *homogeneous charge compression ignition* (HCCI) [80, 128].

The final objective of all the theoretical, experimental, and simulation studies on cycle-to-cycle variations is to be able to predict and control undesired fluctuations. Either physical (adjustable spark advance, optimum ignition systems, and combustion chamber design) or chemical methods (optimum fuel–air ratio, different fuel blends) should be developed and implemented in real engines through the powerful electronic control devices that are available today. This could result in improved driveability, the possibility of using leaner mixtures with reduced fuel consumption, and the reduction of pollutant emissions.

One of the objectives of this monograph is to show how quasi-dimensional simulations can contribute to the better understanding of cyclic variations on spark ignition engines: the main physical and chemical sources of variability, the phenomenology of variations in several engine parameters, and the complex dynamics underlying the evolution of the time series related to the energetics of the engine.

1.4 Organization of the Book

This book is intended to serve as an introduction to the subject of quasi-dimensional modeling of spark ignition engines. We shall develop in a brief but precise way the main physical and chemical laws that are the basis of this kind of models. This will be done in Chap. 2. This chapter also includes a discussion on the basic submodels that are required to build a computer simulation within the framework of quasi-dimensional models. Mathematical models for combustion and empirical relations for the heat transfer processes, frictions, chemical reactions, and details of the fluid properties, will be discussed in this chapter.

In Chap. 3, we describe the different performance parameters susceptible to be measured or computed and present numerical results that can be obtained from quasi-dimensional model simulations to estimate them. We compare the numerical results with experimental observations from real engines and also with predictions from quasi-dimensional simulations performed by other researchers. In addition, a comparison of the results with those obtained from a FTT approach is given. In Chap. 4, we show how the capabilities of quasi-dimensional simulations and the FTT approach can be used together to optimize the performance of a spark ignition engine, in which respect to design and operation parameters.

The last chapter, Chap. 5, is devoted to the analysis of cycle-to-cycle variations using quasi-dimensional simulations. Cyclic variability is achieved by introducing pseudo-random fluctuations in a few essential parameters of a turbulent combustion model. The comparison of the predicted fluctuations and experimental measures is shown. The low requirements of quasi-dimensional simulations in regard to computation times allow to obtain long time series for different variables. We shall concentrate on the evolution of heat release signals with the fuel–air ratio. Statistical analysis of the fluctuations of these series and some nonlinear dynamics techniques are applied to characterize cyclic variability. First return maps, correlation dimension studies, fractal and multifractal analysis, network, and wavelet analysis are analyzed. Finally, a comparative analysis of the different time series arising from different energetic parameters such as heat release, efficiency, and power output will be presented.

Seven appendices are also included. They intend to contribute to the development of quasi-dimensional simulation codes by detailing those aspects that require specific calculations usually difficult to find in the literature in an explicit form. Those appendices include details about the derivation of the basic differential thermodynamic equations, how to manage flow rates, calculations of the flame front area and heat transfer formulas for any possible spark-plug location, and how to state and solve the combustion chemistry, including alternative fuels.

References

1. J. Heywood, *Internal Combustion Engine Fundamentals*, chap. 14 (McGraw-Hill, New York, 1988), pp. 748–819
2. H. Bayraktar, O. Durgun, Energy Sources **25**, 439 (2003)
3. S. Verhelst, C. Sheppard, Energy Convers. Manage. **50**, 1326 (2009)
4. P. Blumberg, J. Kummer, Comb. Sci. Tech. **4**, 73 (1971)
5. J. Heywood, J. Higgins, P. Watts, R. Tabaczynski, SAE Paper 790291 (1979)
6. J. Heywood, *Internal Combustion Engine Fundamentals*, chap. 14 (McGraw-Hill, New York, 1988), pp. 766–768
7. N. Komninos, C. Rakopoulos, Renew. Sustain. Energy Rev. **16**, 1588 (2012)
8. R. Niven, Renew. Sustain Energy Rev. **9**, 535 (2005)
9. K.A. Agarwal, Prog. Energy Combust. Sci. **33**, 233 (2007)
10. S. Verhelst, R. Sierens, Int. J. Hydrogen Energy **32**, 3545 (2007)
11. G. Timilsina, A. Shrestha, Energy **36**, 2055 (2011)
12. Y. Najjar, Open Fuels Energy Sci. J. **2**, 1 (2009)

13. N. Blizard, J. Keck, SAE Paper 740191 (1974)
14. J. Heywood, in *International Symposium COMODIA 94* (1994)
15. S. Hosseini, R. Abdolah, A. Khani, in *Proceedings of the World Congress on Engineering*, vol. II (2008)
16. C. Borgnakke, P. Puzinauskas, Y. Xiao, Spark ignition engine simulation models. Technical report, Department of Mechanical Engineering and Applied Mechanics. University of Michigan. Report No. UM-MEAM-86–35 (1986)
17. G. Beretta, M. Rashidi, J. Keck, Combust. Flame **52**, 217 (1983)
18. D. Bradley, M. Haq, R. Hicks, T. Kitagawa, M. Lawes, C. Sheppard, R. Woolley, Combust. Flame **133**, 415 (2003)
19. J. Heywood, *Internal Combustion Engine Fundamentals*, chap. 14 (McGraw-Hill, New York, 1988), pp. 771–773
20. R. Tabaczynski, C. Ferguson, K. Radhakrishnan, SAE Paper 770647 (1977)
21. J. Keck, J. Heywood, G. Noske, SAE Paper 870164 (1987)
22. J. Heywood, *Internal Combustion Engine Fundamentals*, chap. 14 (McGraw-Hill, New York, 1988), pp. 797–816
23. J. Beauquel, *Computational Fluid Dynamics Modelling: Flow Behaviour in the Combustion Chamber of a Spark Ignition Engine* (VDM Verlag, 2009)
24. H. Versteeg, W. Malalasekera, *An Introduction to Computational Fluid Dynamics. The Finite Volume Methods* (Longman Scientific and Technical, England, 1995)
25. J. Gordon, M. Huleihil, J. Appl. Phys. **72**, 829 (1992)
26. A. Fischer, K. Hoffmann, J. Non-Equilib, Thermodynamics **29**, 9 (2004)
27. I. Novikov, J. Nucl. Energy II **7**, 125 (1958)
28. M. Chambadal, Revue Générale de L'Électricité **67**, 332 (1958)
29. B. Andresen, P. Salamon, R. Berry, J. Chem. Phys. **66**, 1571 (1977)
30. B. Andresen, P. Salamon, R. Berry, Phys. Today **37**, 62 (1984)
31. A. Bejan, *Entropy Generation Through Heat and Fluid Flows* (Wiley, New York, 1982)
32. A. Bejan, J. Appl. Phys. **79**, 1191 (1996)
33. D.P. Sekulic, J. Appl. Phys. **83**, 4561 (1998)
34. M. Moran, Energy **23**, 517 (1998)
35. E. Gyftopoulos, Energy **24**, 1035 (1999)
36. F.L. Curzon, B. Ahlborn, Am. J. Phys. **43**, 22 (1975)
37. A. de Vos, *Thermodynamics of Solar Energy Conversion* (Wiley, New York, 2008)
38. N. Sánchez Salas, S. Velasco, A. Calvo Hernández, Energy Convers. Manage. **43**, 2341 (2002)
39. N. Sánchez Salas, L. López-Palacios, S. Velasco, A. Calvo Hernández, Phys. Rev. E **82**, 051101 (2010)
40. M. Esposito, K. Lindenberg, C. Van der Broeck, Phys. Rev. Lett. **102**, 130602 (2009)
41. M. Esposito, K. Lindenberg, C. Van der Broeck, EPL **85**, 60010 (2009)
42. Z. Tu, J. Phys. A: Math. Theor. **41**, 312003 (2008)
43. L. Chen, F. Sun (eds.), *Advances in Finite-Time Thermodynamics* (Nova Science Publishers, Hauppauge, New York, 2004)
44. A. Durmayaz, O.S. Sogut, B. Sahin, H. Yavuz, Prog. Energy Combust. **30**, 175 (2004)
45. M. Feidt, Entropy **11**, 529 (2009)
46. B. Andresen, Angew. Chem. Int. Ed. **50**, 2690 (2011)
47. C.D. Rakopoulos, E.G. Giakoumis, Prog. Energy Combust. **32**, 2 (2006)
48. R. Ebrahimi, D. Ghanbarian, M. Tadayon, J. Am. Sci. **6**, 27 (2010)
49. M. Gumus, M. Atmaca, T. Yilmaz, Int. J. Energy Res. **33**, 745 (2009)
50. Y. Ge, L. Chen, F. Sun, C. Wu, Int. J. Exergy **2**(3), 274 (2005)
51. Y. Ge, L. Chen, F. Sun, Appl. Energy **85**, 618 (2008)
52. J.C. Lin, S. Hou, Energy Convers. Manage. **49**, 1218 (2008)
53. M. Huleihil, Physics Research International ID 496057 (2011)
54. F. Angulo-Brown, T.D. Navarrete-González, J.A. Rocha-Martínez, in *Recent Advances in Finite-Time Thermodynamics*, ed. by C. Wu, L. Chen, J. Chen (Nova Science Publishers, Commack, New York, 1999)

55. J.A. Rocha-Martínez, T.D. Navarrete-González, C.G. Pavía-Miller, A. Ramírez-Rojas, F. Angulo-Brown, Int. J. Ambient Energy **27**, 181 (2006)
56. M. Huleihil, B. Andresen, J. Appl. Phys. **100**, 114914 (2006)
57. L. Chen, S. Xia, F. Sun, Energy Fuels **24**, 295 (2010)
58. D. Lyon, *in Petroleum based fuels and automotive applications* (I. Mech. E. Conf. Proc., (MEP), London, 1986)
59. C. Stone, A. Brown, P. Beckwith, SAE Paper 960613 (1996)
60. R. Stone, *Introduction to Internal Combustion Engines*, chap. 4 (Macmillan Press LTD., London, 1999), pp. 181–184
61. N. Ozdor, M. Dulger, E. Sher, SAE Paper 940987 (1994)
62. H. Zhang, X. Han, B. Yao, G. Li, Appl. Energy **104**, 992 (2013)
63. J. Heywood, *Internal Combustion Engine Fundamentals*, chap. 9 (McGraw-Hill, New York, 1988), pp. 413–427
64. E. Abdi Aghdam, A.A. Burluka, T. Hattrell, K. Liu, G.W. Sheppard, J. Neumeister, N. Crundwell, SAE Paper 2007–01-0939 (2007)
65. M. Parsl, H. Daneshyar, SAE Paper 892100 (1989)
66. N. Trigui, W. Choi, Y. Guezennec, SAE Paper 962085 (1996)
67. D. Park, P. Sullivan, J. Wallace, SAE Paper 2004–01-1351 (2004)
68. F. Ma, H. Shen, C. Liu, D. Wu, G. Li, D. Jiang, 961969 (1996)
69. G. de Soete, in *Mechanical Engineering Conference Proceeding*, vol. I (Int. Conf. on Combustion Engineering, (MEP), London, 1983)
70. D. Lord, R. Anderson, D. Brehob, Y. Kim, SAE Paper 930463 (1993)
71. R. Winsor, D. Patterson, SAE Paper 730086 (1973)
72. T. Urushihara, T. Murayama, Y. Takagi, K. Lee, SAE Paper 950813 (1995)
73. J. Whitelaw, H. Xu, SAE Paper 950683 (1995)
74. S. Russ, G. Lavoie, W. Dai, SAE Paper 1999–01-3506 (1999)
75. H. Schock, Y. Shen, E. Timm, T. Stuecken, A. Fedewa, P. Keller, SAE Paper 2003–01-1357 (2003)
76. C. Ji, P. Ronney, SAE Paper 2002–01-2736 (2002)
77. F. Zhao, M. Taketomi, K. Nishida, H. Hiroyasu, SAE Paper 940988 (1994)
78. G. Grünefeld, V. Beushausen, P. Andresen, W. Hentschel, SAE Paper 941880 (1994)
79. R. Stone, *Introduction to Internal Combustion Engines*, chap. 4 (Macmillan Press LTD., New York, 1999), pp. 155–164
80. A. Sen, G. Litak, C. Edwards, C. Finney, C. Daw, R. Wagner, Appl. Energy **88**, 1648 (2011)
81. J. Heywood, *Internal Combustion Engine Fundamentals*, chap. 1 (McGraw-Hill, New York, 1988), pp. 37–41
82. A. Schmid, M. Grill, H.J. Berner, M. Bargende, S. Rossa, M. Böttcher, SAE Paper 2009–01-2659 (2009)
83. F. Zhao, M.C. Lai, D. Harrington, Prog. Energy Combust. **25**, 437 (1999)
84. K. Watanabe, S. Ito, T. Tsurushima, SAE Paper 2010–01-0544 (2010)
85. A. Ibrahim, S. Bari, Fuel **87**, 1824 (2008)
86. A. Ibrahim, S. Bari, Energy Convers. Manage. **50**, 3129 (2009)
87. R. Stone, *Introduction to Internal Combustion Engines*, chap. 4 (Macmillan Press LTD., New York, 1999), pp. 185–191
88. G. Kalghatgi, SAE Paper 870163 (1987)
89. L. Camilli, J. Gonnella, T. Jacobs, SAE Paper 2012–04-16 (2012)
90. A. Alkidas, Energy Convers. Manage. **48**, 2751 (2007)
91. I. Altin, A. Bilgin, Energy Convers. Manage. **50**, 1902 (2009)
92. M. Rashidi, Combust. Flame **42**, 111 (1981)
93. C.S. Daw, C.E.A. Finney, J.B. Green, M.B. Kennel, J.F. Thomas, F.T. Connolly, SAE Paper 962086 (1996)
94. C.S. Daw, M.B. Kennel, C.E.A. Finney, F.T. Connolly, Phys. Rev. E **57**, 2811 (1998)
95. E. Pariotis, G. Kosmadakis, C. Rakopoulos, Energy Convers. Manage. **60**, 45 (2012)
96. C. Rakopoulos, G. Kosmadakis, A. Dimaratos, E. Pariotis, Appl. Energy **88**, 111 (2011)

97. E. Galloni, Appl. Therm. Eng. **29**, 1131 (2009)
98. L. Thobois, G. Rymer, T. Soulères, T. Poinsot, SAE Paper 2004–01-1854 (2004)
99. O. Vermorel, S. Richard, O. Colin, C. Angelberger, A. Benkenida, in *International Multidimensional Engine Modeling User's Group Meeting* (2007)
100. C. Lacour, C. Pera, B. Enaux, O. Vermorel, C. Angelberger, T. Poinsot, in *European Combustion Meeting* (2009)
101. R. Stone, *Introduction to Internal Combustion Engines*, chap. 3 (Macmillan Press LTD., New York, 1999), pp. 109–113
102. S. Kumar, M. De-Zylva, H. Waston, SAE Paper 912454 (1991)
103. F. Shen, P. Hinze, J.B. Heywood, SAE Paper 961187 (1996)
104. P. Curto-Risso, A. Medina, A. Calvo Hernández, L. Guzmán-Vargas, F. Angulo-Brown, Appl. Energy **88**, 1557 (2011)
105. P.L. Curto-Risso, A. Medina, A. Calvo Hernández, L. Guzmán-Vargas, F. Angulo-Brown, Physica A **389**, 5662 (2010)
106. P.L. Curto-Risso, A. Medina, A. Calvo Hernández, in *24th International Conference on Efficiency, Cost, Optimization, Simulation and Environmental Impact* (Novi Sad, Serbia, 2011)
107. H. Kantz, T. Schreiber, *Non-Linear Time Serie Analysis* (Cambridge University Press, Cambridge, 2004)
108. P.C. Ivanov, L. Amaral, A. Goldberger, S. Havlin, M. Rosenblum, Z. Stuzik, H. Stanley, Nature **399**, 461 (1999)
109. A. Goldberger, L. Amaral, J. Hausdorff, P.C. Ivanov, C.K. Peng, H. Stanley, Proc. Natl. Acad. Sci. U.S.A. **99**(Suppl 1), 2466 (2002)
110. M. Costa, A.L. Goldberger, C.K. Peng, Phys. Rev. Lett. **89**, 068102 (2002)
111. L. Telesca, V. Lapenna, F. Vallianatos, Phys. Earth Planet. Inter. **131**(1), 63 (2002). doi:10.1016/S0031-9201(02)00014-6
112. C. Letelier, S. Meunier, G. Gouesbet, F. Neveu, T. Duverger, B. Cousyn, SAE Paper 971640 (1997)
113. J.B.J. Green, C.S. Daw, J.S. Armfield, R.M. Wagner, J.A. Drallmeier, M.B. Kennel, P. Durbetaki, SAE Paper 1999–01-0221 (1999)
114. C.S. Daw, C.E.A. Finney, M.B. Kennel, Phys. Rev. E **62**, 1912 (2000)
115. D. Scholl, S. Russ, SAE Paper 1999–01-3513 (1999)
116. A. Sen, G. Litak, T. Kaminski, M. Wendeker, Chaos **18**, 033115 (2008)
117. A. Sen, R. Longwic, G. Litak, K. Gorski, Mech. Syst. Signal Process. **22**, 362 (2008)
118. A. Sen, G. Litak, C. Finney, C. Daw, R. Wagner, Appl. Energy **87**, 1736 (2010)
119. A. Sen, G. Litak, B. Yao, G. Li, Appl. Therm. Eng. **30**, 776 (2010)
120. A. Sen, S. Ash, B. Huang, Z. Huang, Appl. Therm. Eng. **31**, 2247 (2011)
121. A. Sen, J. Zheng, Z. Huang, Appl. Energy **88**, 2324 (2011)
122. G. Litak, T. Kaminski, J. Czarnigowski, D. Zukowski, M. Wendeker, Meccanica **42**, 423 (2007)
123. G. Litak, T. Kaminski, R. Rusinek, J. Czarnigowski, M. Wendeker, Chaos Solitons Fractals **35**, 578 (2008)
124. T. Kaminski, M. Wendeker, K. Urbanowicz, G. Litak, Chaos **14**, 461 (2004)
125. G. Litak, R. Taccani, R. Radu, K. Urbanowicz, J.A. Holyst, M. Wendeker, A. Giadrossi, Chaos Solitons Fractals **23**, 1695 (2005)
126. G. Litak, T. Kaminski, J. Czarnigowski, A.K. Sen, M. Wendeker, Meccanica **44**, 1 (2009)
127. G. Li, B. Yao, Appl. Therm. Eng. **28**, 611 (2008)
128. R. Maurya, A. Agarwal, Appl. Energy **88**, 1153 (2011)
129. S. Wang, C. Ji, Int. J. Hydrogen Energy **37**, 1112 (2012)
130. M. Ceviz, F. Yüksel, Renew. Energy **31**, 1950 (2006)

Chapter 2
Physical Laws and Model Structure of Simulations

2.1 Basic Mechanical Equations

The main components of an internal combustion engine are a cylinder which acts as the combustion chamber, a piston, and a crankshaft. The force generated from burning of the combustible mixture is transmitted from the piston to the crankshaft through a connecting rod. Using Newton's second law, the rotational motion of the crankshaft can be described by the equation[1]:

$$I\dot{\omega} = M_b - M_{\text{ext}} \tag{2.1}$$

where I denotes the *moment of inertia of the crankshaft*, ω is the *angular speed*, M_b is the torque that the connecting rod exerts on the crankshaft, and M_{ext} is the torque associated with the external load. Figure 2.1 is a schematic diagram of the cylinder-piston-crankshaft system. The volume occupied by the gases in the cylinder is minimum when the piston reaches the *top center* (TC), and is maximum when the piston is at the *bottom center* (BC).

With the notation used in Fig. 2.1, the torque, $\mathbf{M}_b$, is given by $\mathbf{M}_b = \mathbf{a} \times \mathbf{F}_\beta$, where $\mathbf{F}_\beta$ is the component of the force exerted by the connecting rod on the crankshaft in the direction of the connecting rod, and a is the *crank radius*. If F is the vertical component of the net force on the piston, $F_\beta = F / \cos \beta$. So

$$M_b = aF\frac{\sin(\varphi + \beta)}{\cos \beta} \tag{2.2}$$

Since $a \sin \varphi = \ell \sin \beta$, it follows that $\sin \beta = f \sin \varphi$, and $\cos \beta = (1 - f^2 \sin^2 \varphi)^{1/2}$, where f is the ratio of the crank radius to the connecting rod length, $f = a/\ell$, and φ is the angle describing the instantaneous position of the piston. This angle is usually called the *crank angle*. After some calculus M_b is given

---

[1] A single dot denotes the first derivative with respect to time and two dots the second derivative.

A. Medina et al., *Quasi-Dimensional Simulation of Spark Ignition Engines*,
DOI: 10.1007/978-1-4471-5289-7_2, © Springer-Verlag London 2014

Fig. 2.1 Notation and geometrical parameters of the
crankshaft-piston system in a
reciprocating engine

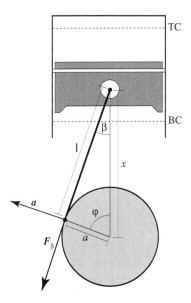

by:

$$M_b = aF \left[\sin \varphi + \frac{f}{2} \frac{\sin 2\varphi}{(1 - f^2 \sin^2 \varphi)^{1/2}} \right] \tag{2.3}$$

To calculate F, we balance the forces acting on the piston[2]:

$$F_{\text{gas}} - F_{\text{fric}} - F + m_p g = m_p \ddot{x} \tag{2.4}$$

In this equation, F_{gas} represents the force exerted by the gas mixture inside the cylinder on the piston head, F_{fric} is the *friction force*, m_p is the piston mass, and $x(t)$ is the position of the piston at any time. This position, $x(t)$, and its time derivatives can be expressed in terms of the angle φ from this geometrical relation:

$$x = a \left[1 - \cos \varphi + \frac{1}{f} \left(1 - \sqrt{1 - f^2 \sin^2 \varphi} \right) \right] \tag{2.5}$$

By substituting the second derivative of $x(t)$ in Eq. (2.4), considering that $\varphi = \varphi(t)$, and using it in Eqs. (2.3) and (2.1), it is straightforward to obtain an expression for the time evolution of the crank angle, φ:

$$\ddot{\varphi} = \frac{a\xi_1 \left(F_{\text{gas}} - F_{\text{fric}} \right) - am_p \xi_1 \left(a\xi_2 \dot{\varphi}^2 - g \right) - M_{\text{ext}}}{I + a^2 m_p \xi_1^2} \tag{2.6}$$

---

[2] The pressure exerted by the oil in the crankcase over the bottom part of the piston in taken as negligible.

where ξ_1 and ξ_2 are geometrical functions depending on the crank angle, φ, and f:

$$\xi_1(\varphi, f) = \sin\varphi + f\frac{\sin\varphi\cos\varphi}{\sqrt{1 - f^2\sin^2\varphi}} \tag{2.7}$$

$$\xi_2(\varphi, f) = \cos\varphi + f\frac{\cos(2\varphi)}{\sqrt{1 - f^2\sin^2\varphi}} + \frac{f^3}{4}\frac{\sin^2(2\varphi)}{\left(1 - f^2\sin^2\varphi\right)^{\frac{3}{2}}} \tag{2.8}$$

In the following sections, we shall go over the calculations of F_{gas} and F_{fric} by means of computer simulation models. F_{gas} is related to the evolution of the pressure inside the combustion chamber and can be estimated from the conservation of energy (first law of thermodynamics) and the equation of state of the gas mixture. Friction forces are usually computed from phenomenological models that can incorporate a more or less detailed description of the possible mechanisms involved in such losses.

To solve the thermodynamic equations of the system, which will be carried out in Sect. 2.2, it is necessary to derive an equation for the variation of the cylinder volume with time or the crank angle. In order to do that, we consider the following definitions (see Fig. 2.1):

- B: *cylinder bore.*
- A_{piston}: *piston area*, $A_{\text{piston}} = \pi B^2/4$.
- L: *stroke* (two times the crank radius, $L = 2a$) .
- V_0: minimum cylinder volume (*clearance volume*). Cylinder volume when the piston is at the top dead center (TC).
- V_{td}: totally displaced or *swept volume*, $V_{td} = A_{\text{piston}} L$.
- $V_d(t)$: *displaced volume* at any time instant, $V_d(t) = A_{\text{piston}} x(t)$.
- r: *compression ratio.* It is the ratio between the maximum and minimum cylinder volumes, $r = (V_{td} + V_0)/V_0$.

Realistic intervals for the numerical values of these parameters in different spark ignition engines, as well as their sample values can be found in several internal combustion engine textbooks [1–3].

With the definitions given above, the total volume, V, at any instant is given by $V = V_0 + V_d = V_0 + A_{\text{piston}} x$. Using $r = \dfrac{V_{td}}{V_0} + 1$ and $A_{\text{piston}} = (r-1)V_0/L$,

$$V = V_0 + (r-1)\frac{V_0}{L}x = V_0\left[1 + (r-1)\frac{x}{2a}\right] \tag{2.9}$$

In terms of the crank position:

$$V = V_0\left[1 + \frac{1}{2}(r-1)\left(\frac{1}{f} + 1 - \cos\varphi - \sqrt{\frac{1}{f^2} - \sin^2\varphi}\right)\right] \tag{2.10}$$

Its time derivative is given by:

$$\dot{V} = \frac{V_0}{2}(r-1)\left(\sin\varphi + \frac{\sin\varphi\cos\varphi}{\sqrt{\frac{1}{f^2} - \sin^2\varphi}}\right) \qquad \omega = \frac{V_0}{2}(r-1)\xi_1(\varphi)\omega \qquad (2.11)$$

where $\omega = \dot{\varphi}$ is the angular speed. This purely kinematic equation is necessary to estimate the friction forces acting on the piston (Sect. 2.3.3), and also to solve the thermodynamic differential equations. There are two cases when one is trying to reproduce the evolution of the engine.

(i) The first one arises when the angular speed, ω, is a fixed input datum. In that case, the mechanical equation, Eq. (2.6), is unnecessary. The dynamics of the system is fully determined by the pressure and temperature equations (that will be described in the next section) and the kinematic equation, Eq. (2.11).

(ii) In the second case, the fixed parameter is M_{ext}, so ω acts as a variable. In this situation, Eq. (2.6), is coupled the thermodynamic equations and Eq. (2.11), so all the equations have to be simultaneously solved.

These equations should be supplemented by the equations for temperature and pressure for the gas mixture inside the cylinder (which is analyzed in the next section), and the mechanical equation, Eq. (2.6). But the latter is unnecessary if the objective is to follow the time or crank angle evolution of the system at a fixed angular velocity. On the other hand, if the external moment, M_{ext}, constitutes an input datum, then Eq. (2.6) should be coupled to the set of thermodynamic differential equations.

2.2 Basic Thermodynamic Equations

In a four-stroke engine, the four strokes correspond to two revolutions of the crankshaft. Figure 2.2 illustrates the different strokes and depicts the evolution of pressure inside the cylinder with respect to the crank angle. The four strokes are:

1. *Intake stroke*: the piston moves from the top center (TC) to the bottom center (BC) introducing fresh mixture into the cylinder. To increase the inducted mass, usually the inlet valve opens shortly before the stroke starts and closes after it ends. Consequently, there exists an *overlap* period when both valves are simultaneously open (see the light shadowed region in Fig. 2.2).

2. *Compression stroke*: all the valves are closed and the mixture is compressed to a lower volume. Close to the end of this stroke, a spark initiates combustion, making pressure to increase rapidly (dark shadowed region in Fig. 2.2).

3. *Power stroke*: it starts with the piston at TC and the gases in the combustion chamber at high temperature and high pressure pushing the piston down and making the crank rotate. Much work is obtained from this stroke in comparison to the work required for compression. Before the piston reaches BC, the exhaust valve opens and the pressure drops.

4. *Exhaust stroke*: the remaining burned gases leave the cylinder. As the piston approaches TC, the inlet valve opens and the cycle begins again.

Fig. 2.2 Evolution of the
pressure inside the cylinder
of a four-stroke spark-ignition
engine as a function of the
crank angle, φ. The top center
(TC) is taken as 0° (end of
compression). *TC* Top center;
BC Bottom center; *IVO* Intake
valve opens; *IVC* Intake valve
closes; *EVO* Exhaust valve
open; *EVC* Exhaust valve
closes

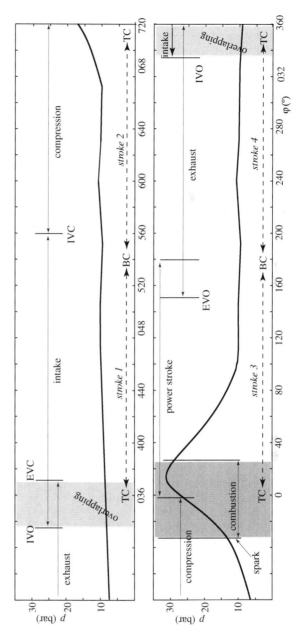

Basic thermodynamic equations stated in a differential form are obtained from
the first law of thermodynamics for open systems and the ideal gas law. The cylinder interior is considered as the control volume. In this section, we formulate the

differential equations for temperature and pressure in each of the four strokes mentioned above. Details about the derivation of these equations from the first law of thermodynamics can be found in Appendix A.

2.2.1 Intake and Exhaust

The intake system of a spark ignition engine consists of an intake manifold, an air filter, fuel injector, and throttle or throttle with individual fuel injectors in each intake port. Pressure losses can occur in each of these elements and also at the intake port and valve. The exhaust system consists of an exhaust manifold, exhaust pipe, a catalytic device, and a muffler. From a thermodynamic standpoint, the system is open during both strokes and the working fluid can be a mixture of fuel, air, and combustion products.

If we denote by $\dot{m}_{in}h_{in}$, the enthalpy changes associated with intake, then the energy conservation and the ideal gas equation of state enables us to write (except during the overlap period, see Appendix A)[3]:

$$\dot{T}_u = \frac{\dot{Q}_u + \dot{m}_u(h_{in} - h_u) + V_u\dot{p}}{m_u c_{p,u}} \tag{2.12}$$

where $\dot{T}_u$ is the temperature of the unburned gases, $\dot{Q}_u$ is the heat transfer from the gas mixture to the cylinder walls during intake, and $c_{p,u}$ is the specific heat at constant pressure of the mixture. It was assumed that $\dot{m}_u = \dot{m}_{in}$.

The only difference in the case of exhaust is the composition of the working fluid (burned gases) and the enthalpy variations that are now denoted with the subscript ex:

$$\dot{T}_b = \frac{\dot{Q}_b + \dot{m}_b(h_{ex} - h_b) + V_b\dot{p}}{m_b c_{p,b}} \tag{2.13}$$

By differentiating the ideal gas equation of state, either for burned or unburned gases, it is easy to obtain the differential equations for pressure during intake or exhaust:

$$\dot{p} = \gamma_u p \left(\frac{\dot{m}_u}{m_u} - \frac{\dot{V}_u}{V_u}\right) + \frac{R_u}{V_u c_{v,u}}\left[\dot{Q}_u + \dot{m}_u(h_{in} - h_u)\right] \tag{2.14}$$

$$\dot{p} = \gamma_b p \left(\frac{\dot{m}_b}{m_b} - \frac{\dot{V}_b}{V_b}\right) + \frac{R_b}{V_b c_{v,b}}\dot{Q}_b \tag{2.15}$$

---

[3] In this and the following equations, the subscript u refers to the gas mixture before combustion. It is composed of fresh air and fuel, but also of reaction products from the previous cycle. The subscript b refers to the gas mixture after combustion, i.e., composed of reaction products. As it will be explained later, both mixtures coexist during combustion in the cylinder. This is one of the main identifying characteristics of two-zone simulations.

where $\gamma_u = c_{p,u}/c_{v,u}$ and $\dot{m}_b(h_{ex} - h_b) \simeq 0$, so the pressure variations during exhaust associated with temperature variations arise only from heat losses through the cylinder walls.[4] Details of the mass fluxes during intake and exhaust and possible valve geometries are given in Appendix B. The initial values for pressure and temperature required to solve these equations are those at the intake and exhaust ducts that should be considered as input data.

It is usual to lengthen the valve open periods beyond intake and exhaust strokes in order to improve filling and emptying of the cylinders. The exhaust process usually begins anywhere between 40° and 60° before BC. Up to BC, gases are discharged due to the pressure difference between the cylinder and the exhaust system. After BC, cylinder is emptied by the piston during its motion toward TC. Typically, the exhaust valve closes 15° to 30° after TC, and the inlet valve opens between 10° and 20° before TC. So both valves are open during an *overlap period*, and depending on the relative pressures it is possible that some of the exhaust gases may flow through the intake manifold and then return to the cylinder with the fresh mixture. Overlap is especially advantageous at high speeds and in high performance engines [4, 5]. The general differential equations for temperature and pressure derived in Appendix A (Eqs. (A.12) and (A.13), respectively) are valid during the overlap period, but in order to fix the sign of the mass fluxes, it is necessary to analyze the relative pressures inside the cylinder and those at exhaust or intake. These specific situations are detailed in Appendix B.

2.2.2 Compression and Expansion

The system is closed during the compression and expansion strokes, so the mass fluxes through the boundaries are zero during these strokes. The mass inside the cylinder during compression is that trapped during intake. Its chemical composition does not change up to the combustion stage and its thermal properties are characterized by an adiabatic coefficient, γ_u, and specific heat at constant pressure, $c_{p,u}$. Similarly, after combustion the chemical composition of the gas mixture is altered, and so its adiabatic coefficient, γ_b, and the specific heat, $c_{p,u}$, change.

The differential equations for temperature and pressure during the compression stroke are:

$$\dot{T}_u = \frac{\dot{Q}_u + V_u \dot{p}}{m_u c_{p,u}} \tag{2.16}$$

$$\dot{p} = \frac{1}{V_u} \left[(\gamma_u - 1) \dot{Q}_u - \gamma_u p \dot{V}_u \right] \tag{2.17}$$

---

[4] It is considered that the gas mixture leaving the control volume is essentially formed by burned gases: all the fuel was burned out and the fresh mixture leaving out during the overlap period (see below) is negligible. There are no appreciable temperature changes from inside the cylinder to the exhaust duct, so $\dot{m}_b(h_{ex} - h_b) \simeq 0$.

For the power stroke, these equations are:

$$\dot{T}_b = \frac{\dot{Q}_b + V_b \dot{p}}{m_b c_{p,b}} \tag{2.18}$$

$$\dot{p} = \frac{1}{V_b} \left[(\gamma_b - 1) \dot{Q}_b - \gamma_b p \dot{V}_b \right] \tag{2.19}$$

Both these sets of equations become identical if we interchange the subscripts u and b, but the thermal properties of the gases are different before and after combustion.

2.2.3 Pressure and Temperature Evolution During Combustion

If a two-zone scheme is considered, both the unburned and burned gases coexist inside the combustion chamber during combustion. They have different temperatures, T_u and T_b, and the same pressure, p. The propagating flame separates the two zones. Experimental studies show that for gasoline engines, the flame is reasonably well defined and that it has a nearly spherical shape with a relatively thin structure. Flame propagation is associated to the rate at which unburned gases transform into burned gases. This implies the nonexistence of heat transfer between the two zones, except for radiation which is not very significant in gasoline engines. In general, it is not a wrong assumption to consider a fixed origin for the flame kernel, coinciding with the spark plug position, although strictly the flame kernel can be a point whose position changes with time, especially when swirl effects are considered.

In this section, we formulate the basic equations to describe the thermodynamic evolution of the system. Explicit combustion models to evaluate the rate at which unburned gases transform into burned gases causing the flame to propagate will be analyzed in Sect. 2.3.1. Following the procedure of [6], we consider the basic equations for the burned and unburned gases, mass and energy conservation, and ideal gas law.

- Unburned gases.

$$\dot{m}_u = -\dot{m}_b \tag{2.20}$$

$$\dot{T}_u = \frac{\dot{Q}_u + V_u \dot{p}}{m_u c_{p,u}} \tag{2.21}$$

$$p V_u = m_u R_u T_u \tag{2.22}$$

- Burned gases.

$$\dot{m}_b = \text{combustion rate} \tag{2.23}$$

$$\dot{T}_b = \frac{1}{m_b c_{p,b}} \left[\dot{m}_b (h_u - h_b) + \dot{Q}_b + V_b \dot{p} \right] \tag{2.24}$$

$$p V_b = m_b R_b T_b \tag{2.25}$$

A constraint for the total volume, V, should be added to these equations:

$$\dot{V} = \dot{V}_u + \dot{V}_b \tag{2.26}$$

Equations (2.21) and (2.24) enable us to follow the evolution of the temperature in the two zones, once the combustion rate, $\dot{m}_b$, is known. In Sect. 2.3.1, we shall use explicit combustion models to compute $\dot{m}_b$. An equation for pressure variations can be obtained by differentiating the ideal gas laws, substituting in the constraint for volume, and eliminating the temperature changes by means of the equations for the energy conservation. This leads to:

$$\dot{p} = \left[p \left(\frac{\dot{m}_b}{\rho_b} + \frac{\dot{m}_u}{\rho_u} - \dot{V} \right) + \frac{\dot{Q}_u R_u}{c_{p,u}} + \right.$$
$$\left. + \left(\dot{Q}_b + \dot{m}_b (h_u - h_b) \right) \frac{R_b}{c_{p,b}} \right] \left(V - \frac{V_u R_u}{c_{p,u}} - \frac{V_b R_b}{c_{p,b}} \right)^{-1} \tag{2.27}$$

The initial values for T_u and p are obtained through the corresponding differential equations for the compression stage which are given by Eqs. (2.16) and (2.17). The initial value of T_b is taken as the *adiabatic flame temperature* at constant pressure [7, 8] which is calculated by solving the equation for the chemical equilibrium (see Sect. 2.3.4, and Appendices D and E).

2.3 Additional Submodels

The resolution of the thermodynamic set of differential equations for each stroke requires, apart from the corresponding initial conditions in each case, submodels about the combustion process (including chemical reactions), the heat transfer from the gas mixture to the cylinder walls, frictions, and the thermal properties of the gas mixture. The purpose of this section is to analyze these basic submodels.

2.3.1 Combustion Models

In a spark ignition engine, combustion begins by the action of a spark discharge. Experimental work has demonstrated that at the start of combustion, the flame is a smooth surface roughly spherical kernel about 1 mm in diameter and grows for the next degrees with an approximately spherical shape [7, 9, 10]. This period is called *initial burning phase* or *flame-development region*. During this stage a small fraction of the mass is burned and the burning speed is close to the *laminar flame speed*. Several crank degrees after, the interaction of the flame with the turbulent gas flow turns out a highly wrinkled and convoluted outer surface of the flame [9, 10]. Burning speed now becomes *turbulent flame speed*. This stage is known as *fast burning phase* or *rapid burning region*. After the end of the flame propagation, it is experimentally known that not all the fuel mass is burned. This last stage is the *final burning phase*.

Since the combustion process directly affects engine performance and emissions, the development of an appropriate combustion model is crucial in any computer simulation of internal combustion engines. The combustion process determines the amount of heat release and its timing, and this, in turn, influences all the other phenomena during engine operation. Specifically, we will show in Chap. 5 that a combustion model can be used effectively to describe the cycle-to-cycle variability in engine operation.

The rate of combustion depends strongly on the flow field in the combustion chamber. This flow field is always turbulent instead of laminar in real operating conditions. Turbulent flames evolve at speeds which are one order of magnitude higher than laminar flames. As stated by Borgnakke [6] and Heywood [11], although much theoretical and experimental work is being developed in order to complete an analytical theory of turbulent combustion (of which the statistical theory of turbulence is a major ingredient), the practical need to simulate and develop engines has led to pragmatic approaches or models that can be incorporated into computer codes without excessive difficulties.

Experimental work over the past few years has revealed some basic peculiarities in the combustion processes associated to spark ignition engines. These characteristics can be very useful to develop and use new or existing approaches. They are summarized as follows [6, 11, 12]:

1. In spark ignition engines, the charge (formed by fuel, air, and residual gases) is gaseous and is approximately uniformly premixed.
2. Combustion does not take place at a constant volume, but it is complicated by the piston motion.
3. The volume of the oxidation zone where the main chemical reactions occur is small, and provided that *swirl* and *squish*[5] are not very strong, the averaged *flame front*[6] can be approximated as a portion of the surface of a sphere. In consequence, for a given combustion chamber geometry and spark plug position, the spherical

---

[5] *Swirl*, as defined by Ferguson [13], is the rotational flow within the cylinder about its axis. It is associated to the existence of a nonzero angular momentum of the incoming flow. It is used in the design of some gasoline engines with the aim to boost a fast burn and in diesel engines to provoke a rapid mixing between fuel and air. *Squish* is the radial gas motion that takes place at the end of the compression stroke when the piston and the cylinder heads approach each other closely. It is especially associated to direct injection engines.

[6] The *flame front* is considered as the forward boundary of the reacting zone.

burning area, its volume, and the combustion chamber inner area wetted by the burned gases, can be calculated as functions of the flame radius (see Appendices C and D). The heat transfer to the cylinder walls is basically produced in the burned gas zone, so its calculation is associated with the estimation of the area wetted by the burned gases.

4. During combustion, the components in the combustion chamber can be separated into two zones consisting of burned and unburned gases. This is an important feature in the thermodynamic analysis of the combustion process.

A recent critical review of the advancements made in gasoline engines for the reduction of fuel consumption and engine-out emissions is given by Alkidas [14].

2.3.1.1 Zero-Dimensional Models

Probably, the simplest way to simulate combustion in a spark ignition engine is by considering an empirical relationship between the combustion rate and the time or the crank angle through some parameters, such as those defining the start of combustion, φ_0, and its duration, $\Delta\varphi$. These models, although they lack a fundamental basis, are simple and can be readily implemented in simulations. The particular values of the parameters required are calculated from experimental observations, and are associated with specific combustion chamber geometries and combustion chemistry. Consequently, they have limited utility for predicting performance parameters under changing conditions. One of the widest employed models uses a *Wiebe function* to correlate the combustion rate and crank angle,[7]

$$x_b(\varphi) = 1 - \exp\left[-a\left(\frac{\varphi - \varphi_0}{\Delta\varphi}\right)^{m+1}\right] \tag{2.28}$$

where x_b is the mass fraction of burned gases. So $\Delta\varphi$ covers the crank angle interval from $x_b = 0$ up to $x_b \simeq 1$. The parameters a and m are empirical, and are usually taken as $a = 5$ and $m = 2$ [11, 18]. Several works in the literature provide empirical relations to calculate $\Delta\varphi$ in terms of the rotational speed, fuel ratio, pressure ratio, or the delay between the spark ignition and the beginning of combustion [17, 19–21]. Particularly, Soylu and van Gerpen [22] recently developed empirical relations for a and $\Delta\varphi$ as functions of the fuel–air ratio for a natural gas engine. This type of models, in which only a control volume is considered, even during combustion, and that are founded on pure empirical basis are called *single-zone* models, as it was explained in Sect. 1.1.

---

[7] A cosine burning law is also usual in the literature [6, 15–17]:

$$x_b(\varphi) = \frac{1}{2}\left[1 - \cos\left(\frac{\pi(\varphi - \varphi_0)}{\Delta\varphi}\right)\right].$$

Hofmann and Fischer [23] consider another zero-dimensional model in which it is assumed that the flame front speed is constant, so its area initially increases as t^2. The effects on the flame speed of the fuel, combustion chamber geometry, ignition location, etc, are expressed in a single parameter, β_c. The rate for the evolution of the number of moles of unburned gases, $\dot{n}_u$, is given by:

$$\dot{n}_u = -n_{u,0}\frac{\beta_c^3 t^2}{2}e^{-\beta_c t} \tag{2.29}$$

where $n_{u,0}$ is the initial number of moles of the fuel–air mixture. Using this equation, and setting the time to zero at the beginning of combustion, it is possible to express the mole fraction, $\tilde{x}_b$, of burned gases as:

$$\tilde{x}_b = 1 - \frac{1}{2}e^{-\beta_c t}(\beta_c^2 t^2 + 2\beta_c t + 2) \tag{2.30}$$

2.3.1.2 Quasi-Dimensional Models

As stated in Chap. 1, within the quasi-dimensional framework, the mass rate of burned gases is described by means of a single or a set of differential equations, and may depend on the flame area and the flame speed [20, 24–30]. They were developed to bridge the gap between zero- and multi-dimensional models. At modest computational costs they introduce a spatial dependence on the combustion process, so they do relate the model outputs to the combustion chamber geometry and several flow-field parameters. Hereafter, we describe two models that have been widely used during the past few years. Both of them consider that the combustion process is turbulent and the flame front thickness plays a central role.

1. When the *flame front thickness* is of the same order that *Kolmogorov microscale*,[8] flame develops in a *wrinkled laminar flame regime* [31, 32] and combustion is characterized by a *turbulent flame speed*, S_T. It is defined as the velocity at which the mean flame contour propagates into the unburned mixture ahead of the flame [32].[9] It can be expressed as:

$$S_T = \frac{\dot{m}_b}{A_f \rho_u} \tag{2.31}$$

where $\dot{m}_b$ represents the mass of charge burned, ρ_u, the unburned gases density, and, A_f, is the average flame front area considered as approximately spherical. Poulos and Heywood [32, 33] assume Eq. (2.31) as a differential equation giving the evolution of $\dot{m}_b$:

---

[8] In a turbulent flow, Kolmogorov microscale is the eddy size at which molecular viscosity becomes important.

[9] Typical values for the quantities which characterize engine flames can be found in [32].

$$\dot{m}_b = A_f \rho_u S_T \tag{2.32}$$

and that the turbulent flame speed is proportional to the laminar flame speed, S_L,[10] through a phenomenological factor, f [35, 36],

$$S_T = f\, S_L \frac{\rho_u/\rho_b}{(\rho_u/\rho_b - 1)x_b + 1} \tag{2.33}$$

and

$$f = 1 + 0.0018\, N \tag{2.34}$$

where N is the engine speed in rpm, ρ_b the burned gas density, and x_b the mass fraction of burned gases. This model has been satisfactorily used to predict the performance and exhaust emissions of a four-stroke ignition engine fueled with hydrogen-gasoline mixtures and other alternative fuels [35, 37, 38].

2. When the flame front area thickness is larger than Kolmogorov microscale, but smaller than the integral length scale (that characterizes the largest vortex size in the flow) the flame develops in a flamelets-in-eddies or eddy burning regime. During the mid1980s, Keck and coworkers [24, 25, 39, 40] developed a framework that it is usually known as *entrainment* or *eddy burning model*. During combustion, because of turbulence, unburned eddies of *characteristic length*, l_t, are entrained into the flame zone at a velocity $u_t + S_L$, where u_t is the *characteristic speed* (similar to the local turbulence intensity [41]) and S_L the *laminar flame speed* (Fig. 2.3). A characteristic time is defined as: $\tau_b = l_t/S_L$. Thus, the combustion process is described by solving the following system of ordinary differential equations,

$$\dot{m}_e = A_f \rho_u (u_t + S_L)$$
$$\dot{m}_b = A_f \rho_u S_L + \frac{(m_e - m_b)}{\tau_b} \tag{2.35}$$

where m_e denotes the total mass inside the flame front, burned gas and unburned eddies, and A_f is the area of the spherical flame front. The flame front area can be calculated from the average flame front radius, assuming an spherical flame front (Appendix C).

Thus, entrainment combustion models consider two steps: first, unburned mass is entrained by the flame front at a rate given by the equation for $\dot{m}_e$ in Eq. (2.35); second, the unburned turbulent eddies burn in a time that is a function of the eddy size, l_t, and the laminar flame speed, S_L. This is given by the equation for $\dot{m}_b$ in (2.35). During combustion these equations are coupled with the thermodynamic

---

[10] The laminar flame speed, S_L, is defined as the velocity relative to and normal to the flame front, with which unburned gas (locally) moves into the front and is transformed into products under laminar flow conditions [34].

ones, namely Eqs. (2.21), (2.24), and (2.27). The interested reader can find details about these models and many others in the specialized literature [32, 34, 42–44]. In order to solve the equations for the mass rates, Eq. (2.35), two assumptions are considered [30, 41]:

a. When combustion begins, experimental observations indicate an initially quasi-spherical relatively smooth flame kernel [45]. The flame speed is taken as the laminar flame speed, $u_t + S_L \simeq S_L$, and thus a time of the order of τ_b is required to transform it into the turbulent speed.[11] So the equation for the evolution of the total mass inside the flame front, m_e, namely, Eq. (2.35) includes a decreasing exponential term:

$$\dot{m}_e = A_f \rho_u \left[u_t \left(1 - e^{-t/\tau_b} \right) + S_L \right] \tag{2.36}$$

b. Toward the end of combustion, when the volume inside the flame front is approximately the total volume of the chamber ($A_f \rightarrow 0$), the increase of heat losses from the flame due to its proximity to the walls decreases the mass burning rate. Finally, the flame extinguishes mainly due to depletion of the reactants.[12] The burned gas mass rate is expressed as:

$$\frac{\dot{m}_b}{\dot{m}_{b,w}} = \exp \left[\frac{-(t - t_w)}{\tau_b} \right] \tag{2.37}$$

where the subscript w denotes the final burning stage (flame front volume equals chamber volume). The time the flame front takes to reach the cylinder walls is t_w; so Eq. (2.37) holds only for $t > t_w$.

Before performing numerical computation of the differential Eqs. (2.32) or (2.35) for the evolution of masses, some parameters need to be estimated from empirical correlations or from more fundamental models. These parameters are: the flame front area, A_f, (Appendix C contains methods for its evaluation either for centered or noncentered ignition), the laminar flame speed, S_L, the characteristic speed, u_t, and the characteristic length, l_t. They depend on the chemical reactions, type of fuel, fuel–air ratios, the presence of residual gases, the thermodynamic conditions, and some geometric parameters of the cylinder. Next we summarize some empirical correlations that can be found in the combustion literature.

1. For the calculation of the laminar flame speed, S_L, for a given fuel, different relations exist in the combustion literature that express S_L in certain pressure and

---

[11] To have an approximate idea, for $t \simeq 3\tau_b$, the term in the brackets in Eq. (2.36) is around 95 % of $u_t + S_L$. Typical characteristic speeds, u_t, are in the interval $2 - 6$ m/s (except for hydrogen, which are much faster) [25, 46, 47] and laminar flame speeds, S_L, are around one order of magnitude smaller [34] depending on the fuel type, fuel–air ratio, and several other factors.

[12] For practical purposes, in simulations it is usually considered that combustion finishes at the moment the exhaust valve opens, so there could be a small fraction of unburned fuel in the combustion chamber among the residuals.

Fig. 2.3 Schematic representation of the entrainment
model developed by Keck et
al. [24, 25]. The flame front
(that is considered approximately spherical) advances
outward at the laminar flame
speed, S_L. Simultaneously,
unburned mixture crosses the
flame front at a characteristic
speed, u_t, due to turbulent
convection (*inset*). Inside
the flame front, there exist
unburned eddies that have a
typical length scale, l_t

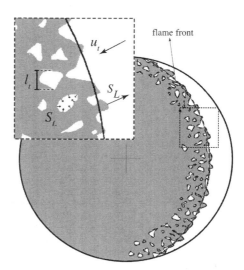

temperature intervals in terms of reference conditions, (T_{ref}, p_{ref}). We will adopt
a widely used example of those relations, although for simulations under specific
conditions or with particular fuels, a literature survey would be necessary.

$$S_L = S_{L,0} \left(\frac{T_u}{T_{ref}}\right)^\alpha \left(\frac{p}{p_{ref}}\right)^\beta \left(1 - 2.06\, y_r^{0.77}\right) \tag{2.38}$$

This correlation described by Heywood [34] explicitly considers the existence of
a fraction of the residual gases, y_r, in the gas mixture (such as CO_2 and N_2), that
causes an appreciable reduction of the laminar burning velocity. The exponents
α and β, and the reference laminar flame speed, $S_{L,0}$, empirically depend on the
fuel–air ratio,[13] ϕ, as:

$$\alpha = 2.18 - 0.8\,(\phi - 1)$$
$$\beta = -0.16 + 0.22\,(\phi - 1)$$
$$S_{L,0} = B_m + B_\phi\,(\phi - \phi_m)^2 \tag{2.39}$$

These relations are adequate for propane, isooctane, and methanol. Here ϕ_m represents the fuel–air ratio corresponding to maximum $S_{L,0}$, denoted by B_m. Table 2.1
lists the parameters ϕ_m, B_m, and B_ϕ required to evaluate $S_{L,0}$ at the reference
conditions, $(T_{ref} = 298$ K, $p_{ref} = 1$ atm$)$ for some fuels.
Another set of correlations for α and β proposed by Rhodes and Keck [50] for
reference gasoline is given by:

---

[13] For brevity we shall use the term fue ratio to refer to the ratio between the masses of fuel and
air and their stoichiometric counterparts (see Eq. (A.3) in Appendix A) that is usually known as
fuel–air equivalence ratio [48].

$$\alpha = 2.4 - 0.271 \, \phi^{3.51}$$
$$\beta = -0.357 + 0.14 \, \phi^{2.77} \tag{2.40}$$

Gulder [51, 52] proposed a different expression for $S_{L,0}$, adequate for ethanol-isooctane blends:

$$S_{L,0} = 0.4658 \, \phi^{-0.326} e^{-4.48(\phi - 1.075)^2} \tag{2.41}$$

taking $\alpha = 1.56$ and $\beta = -0.22$.

2. In order to calculate u_t and l_t, there are other empirical correlations similar to those developed by Keck [24]. For example, Beretta [25] used Eq. (2.35) to different experimental data. The characteristic speed, u_t, depends on the *average inlet gas speed*, U_i, and the ratio between the densities of the unburned gases, and the average density at intake, ρ_u/ρ_i, as:

$$u_t = 0.08 \, U_i \left(\frac{\rho_u}{\rho_i} \right)^{\frac{1}{2}} \tag{2.42}$$

Here

$$U_i = \frac{1}{\Delta\varphi_o} \int_0^{\Delta\varphi_o} c(\varphi) \, d\varphi \tag{2.43}$$

where $\Delta\varphi_o$ is the crank angle interval over which the intake valve is open, and $c(\varphi)$ is the speed of the gas entering the cylinder. Beretta [25] proposes a correlation function for U_i in terms of the *volumetric efficiency*, e_v,[14] the piston area, A_{piston}, the maximum passage area through the intake valve, A_{iv}, and the average piston speed, v_p. This correlation is given by:

$$U_i = e_v \frac{A_{\text{piston}}}{A_{iv}} v_p \tag{2.44}$$

where $v_p = 2\omega a/\pi$, a is the crank radius, and ω is given in rad/s.

3. The characteristic length of eddies, l_t, is obtained from the maximum lift of the intake valve, $L_{v,\max}$, and the ratio ρ_i/ρ_u.

$$l_t = 0.8 \, L_{v,\max} \left(\frac{\rho_i}{\rho_u} \right)^{\frac{3}{4}} \tag{2.45}$$

Figure 2.4 represents the typical S shaped evolution curve for the simulated mass fraction of burned gases, x_b, in terms of the crankshaft angle, φ. The solid curve represents the first cycle characterized by a fast combustion (higher slope in the steepest region) because of the absence of residual gases in the combustion chamber.

---

[14] The *volumetric efficiency*, e_v, is a parameter used to quantify the effectiveness of the intake process. It is the mass that enters the cylinder during intake over the displaced volume $e_v = m_{\text{in}}/(\rho_i V_{dt})$. ρ_i is the density of the mixture evaluated at a reference state; for instance the average conditions on the intake manifold.

Table 2.1 Coefficients required to evaluate the reference laminar flame speed, $S_{L,0}$ (Eq. (2.39)), at ($T_{ref} = 298$ K, $p_{ref} = 1$ atm) [34, 49, 50]

Fuel	ϕ_m	B_m (cm/s)	B_ϕ (cm/s)
Methanol	1.11	36.9	−140.5
Propane	1.08	34.2	−138.7
Isooctane	1.13	26.3	−84.7
Gasoline	1.21	30.5	−54.9

The other curves in the figure (dashed) were obtained after an equilibration period that ensures that the engine has reached a stationary regime. For any of the curves, there are three clear regions (see [53] for details about the usual shape of $x_b(\varphi)$):

1. Initial slow combustion where x_b has a small slope associated with the laminar flame speed, S_L. It is the *flame-development region*, and it lasts up to approximately $x_b = 10\%$.
2. Rapid high slope combustion where the characteristic speed, u_t, is much larger than the laminar speed, S_L (at least one order of magnitude larger). It is the so-called *rapid-burning region*. Approximately, it embraces a crankshaft angle between $x_b = 10\%$ to $x_b = 90\%$.[15]
3. Final slow combustion until burned gases occupy almost the entire volume of the combustion chamber. The flame front does not spread further, but there can be a small amount of unburned fuel. It is the *final burning phase*.

Although limited by their simplicity, the combustion models we have presented above lead to fair agreement when the simulations results are compared with experiments (explicit comparison will be shown in Chap. 3). These models are adequate for *normal combustion*, *i.e.*, the flame starts from a timed spark at an ignition point and spreads continuously outward from the ignition point up to the end of the chamber. Other problems like *abnormal combustion*,[16] inhomogeneity of the turbulence for intricate geometries associated with *swirl* and *squish*, and the relevance of the flow field after intake and compression at the beginning of combustion, will require more detailed multi-dimensional models. Inclusion of these effects will require much more complex and computationally expensive simulations. Knock models susceptible to be implemented in two-zone combustion models are detailed in [56, 57] and swirl effects are analyzed in [58] by means of quasi-dimensional simulations. Excellent

---

[15] These particular numerical values are arbitrary but useful for practical reasons.

[16] Although abnormal combustion comprises several phenomena, probably the most important are *knock* and *surface ignition* [54, 55]. Knock occurs when the mixture appears to ignite and burn ahead of the flame without any external ignition source. There is a rapid release of chemical energy of the gas which provokes pressure waves of considerable amplitude and a particular noise transmitted through the engine structure. Surface ignition is caused by the mixture igniting as consequence of a contact with a hot surface (overheated valve or spark plug) or hot-spot combustion chamber deposits. It may happen before the spark ignites the charge (pre-ignition) or after normal ignition (post-ignition).

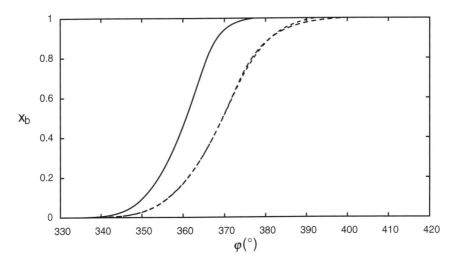

Fig. 2.4 Particular example of simulation results for the burned gases mass fraction, x_b, as a function of the crankshaft angle, φ, for three different cycles at a speed of 109 rad/s and a spark ignition angle of 330°. The first cycle (*solid line*) has a steeper slope in the rapid-burning region because of the absence of residual gases. Fuel is isooctane and x_b was computed with Eq. (2.35)

reviews on this phenomenology are written by Blumberg et al. [16], and Verhelst and Sheppard [45]. Rakopoulos et al. [59] recently investigated the effect of crevice flow on internal combustion engines by means of CFD models. Quasi-dimensional models should be considered as straightforward models (with evident phenomenological foundations), that allow one to incorporate the most relevant physical and chemical mechanisms, and to recover a wide variety of experimental results at a small computational cost.

2.3.2 Heat Transfer Models

The maximum temperature of the burned gases inside the cylinder of a spark ignition engine can reach around 2500 °C, but the temperatures that the materials used to build the cylinder head and walls, and piston are considerably lower. Because of this, an effective cooling system is necessary to ensure proper engine operation.[17] The heat rejected to the coolant depends on several engine variables: speed, load, ignition timing, and fuel–air ratio. Maximum heat transfer occurs during combustion (around 3 MW/m^2), but at other steps the heat flux can be very reduced [62]. In a typical

---

[17] Very briefly, the main reasons for engine cooling are: to get an elevated volumetric efficiency, to avoid combustion problems, and to guarantee proper mechanical operation and reliability. Advanced cooling systems have been analyzed by Robinson et al. [60] and Yoo et al. [61].

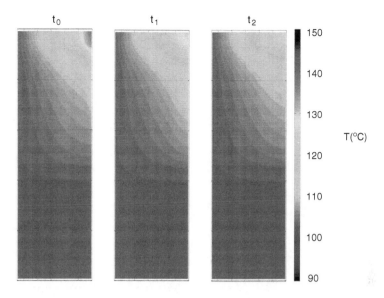

Fig. 2.5 Temperature field for a cylinder wall of an SI engine at three different times, t_0, t_1, and t_2. It was calculated at a constant speed, 1500 rpm, by means of a simplified geometry model, and solving a partial differential equation for heat. The gas is to the *right* of the wall

spark ignition engine, the amount of fuel energy that passes to the coolant can reach 40 % at low loads [63].

The heat flux to the coolant arises from in-cylinder heat transfer and also from the exhaust port. Thermal stresses inside the cylinder should not exceed values that can cause fatigue cracking. The temperatures should not exceed 400 °C for cast iron, and 300 °C for aluminum alloys. The internal surface of the cylinder walls should be below 180 °C to avoid degradation of the lubricating oil film. In addition, spark plug and valves need to be relatively cool to avoid pre-ignition and knock.

The rate of heat transfer has consequences in engine performance and emissions. For a given mass of fuel within the cylinder, high heat transfer rate results in lower combustion temperature and gas pressure, leading to a lower work per cycle. With respect to exhaust, higher exhaust temperature causes further oxidation of combustion products such as CO or HC [16], but also increases the energy that an exhaust turbocharger turbine can recover as work. Friction is directly related to heat transfer. Heat produced by mechanical friction has additionally to be evacuated to the atmosphere by the cooling system. A detailed description of the importance of heat transfer in the design of internal combustion engines can be found in [62, 64].

Because of the transient nature of the system, it is expected a nonuniform temperature field in the cylinder wall. Figure 2.5 depicts as an example the temperature field for a cylinder wall of an SI engine at three different times. It was calculated at a constant speed, 1500 rpm, for a simplified geometrical model by solving the heat diffusion equation inside the sleeve. The results show a small region with high

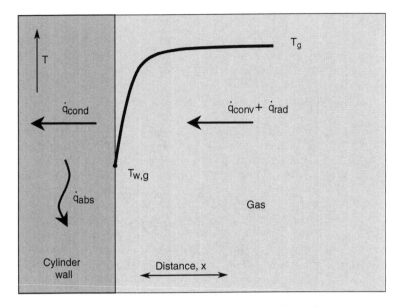

Fig. 2.6 Schematic diagram of the temperature (*solid line*) distribution from the gas inside the cylinder [65, 66]

oscillating temperatures in the upper inner corner of the wall, while in the rest of the volume they remain substantially constant and uniformly distributed. Even though the bulk temperatures of the gases during combustion reach up to 2250 °C in this case, temperatures in the wall do not exceed 140 °C when water with phase change at 100 °C is considered as refrigerant. Most of the volume remains at a temperature below 110 °C, so the highest temperature differences in the wall are close to 30 °C. Therefore, assuming a uniform temperature for the cylinder wall could be a good first approximation for quasi-dimensional models.

From the point of view of engine modeling and simulation, a description of both spatial and temporal variations of the heat transfer coefficients would be required if thermal stresses are important, as for a detailed emissions study [67]. For a global estimation of average heat transfer, simple spatial and time averaged correlations could be adequate [68]. In general, spatial averages can be assumed, but time dependence of heat transfer has to be modeled. However, engine output parameters such as efficiency are not very sensitive to the heat transfer model considered. Differences in the predicted engine performance as obtained from different heat transfer models will be demonstrated in Sect. 3.3.2.

Figure 2.6 depicts a conventional schematic representation of the heat fluxes from the gas inside the cylinder to the coolant through the cylinder walls [65, 66]. The heat transfer rate per unit area from the gas to the inner cylinder walls is composed of two terms, arising from convection and radiation ($\dot{q}_{conv}$ and $\dot{q}_{rad}$), although the radiation term is either considered negligible or subsumed into an effective linear term through an effective heat transfer coefficient:

$$\dot{q} = \dot{q}_{conv} + \dot{q}_{rad} = h_{c,g}(T_g - T_{w,g}) + \varepsilon\sigma(T_g^4 - T_{w,g}^4) \simeq h_{c,g}(T_g - T_{w,g}) \quad (2.46)$$

where $h_{c,g}$ is the *convection (or effective) heat transfer coefficient* at the gas side, ϵ is the *emissivity*, σ the *Stefan-Boltzmann constant*, and $T_{w,g}$ is the temperature of the internal side of cylinder walls. Finally, T_g is the gas instantaneous temperature. The step where more heat is transferred to the walls is during combustion. In the case of two-zone models, it should be distinguished contributions to the heat transfer rate arising from the unburned and the burned gases that are at different temperatures. Detailed calculations for the portion of inner walls area wetted by one or more have to be implemented in order to develop simulations. Appendix D contains specific calculations on this point.

Actually, these coefficients may vary with space and time and their accurate evaluation is a complex task that could be achieved, for instance, by means of Computational Fluid Dynamics (CFD) simulations predicting the in-cylinder gas motion [69, 70]. At another level, phenomenological models derived from experimental measurements can provide approximation to the heat transfer coefficients. We discuss below four models that have been widely used in quasi-dimensional simulations of spark ignition engines [71–74] to evaluate the heat transfer from the gas inside the cylinder to the inner cylinder surface [Eq. (2.46)]. From the assumption of an average cylinder internal wall temperature, $T_{w,g}$, and the selection of any of these models, the heat transfer from the gas to the cylinder walls can be obtained in order to simulate the evolution of the engine.

1. One of the earliest empirical models was developed by Eichelberg [72, 75]. In this phenomenological correlation, $\dot{q}_s$ is a function of several time or crank angle dependent parameters.

$$\dot{q}_s(t) = \frac{\dot{Q}_s(t)}{A_s(t)} = 2.43\, v_P^{1/3} \left[p(t)T_g(t) \right]^{1/2} \left[T_g(t) - T_{w,g} \right] \quad \left(W/m^2 \right) \quad (2.47)$$

where:

$\dot{q}_s(t)$ is the instantaneous heat flow rate through cylinder walls per unit area (W/m^2).
v_P is the mean piston velocity (m/s).
$p(t)$ is the instantaneous pressure inside the cylinder (bar).
$T_g(t)$ is the instantaneous bulk gas temperature (K).
$T_{w,g}$ is the mean temperature of the inner surface of the cylinder (K).
$A_s(t)$ is the instantaneous surface area (m^2).

As mentioned by Stone [75], the main criticism to this model arises from its inconsistent dimensionality and the necessity to use particular units, unless the constant is recalculated.

2. Annand and Ma [76, 77] proposed a model depending on the Reynolds number that explicitly considers radiative heat flux. This model can also be used for compression ignition engines.

$$\dot{q}_s(t) = c \frac{k}{B} \text{Re}^b \left[T_g(t) - T_{w,g} \right] + d \left[T_g^4(t) - T_{w,g}^4 \right] \tag{2.48}$$

where:

k is the gas thermal conductivity [W/(m K)].
Re is the Reynolds number.
B is the cylinder bore (m).
ρ is the gas density (kg/m^3).
μ is the gas dynamic viscosity [kg/(m s)].

The value of the constants b, c, and d was suggested by Watson and Janota [78]:

$b = 0.7$, for a compression ignition engine.
$0.25 < c < 0.8$, it depends on the intensity of charge motion and engine design, increasing with it.
$d = 0.075\sigma$, for a spark ignition engine, only during combustion. Otherwise $d = 0$, i.e., the radiation term is negligible.

3. Woschni [63, 71] proposed a model for the convective heat transfer coefficient, $h_{c,g} = \dot{q}_s / (T_g - T_{w,g})$:

$$h_{c,g}(t) = 129.8 \, p^{0.8} u(t)^{0.8} B^{-0.2} T_g(t)^{-0.55} \quad \left[\text{W}/(\text{m}^2 \text{ K}) \right] \tag{2.49}$$

where u is a characteristic velocity,

$$u(t) = C_1 v_P + C_2 \frac{V_{dt} T_r}{p_r V_r} [p(t) - p_m] \quad (\text{m/s}) \tag{2.50}$$

Here V_{dt} is the swept volume (m^3). The symbols p_r, V_r, T_r are respectively the pressure, volume, and temperature at a reference condition, such as the inlet valve closure, and p_m is the motoring pressure (bar).
Watson and Janota [78] considered compression and expansion as polytropic processes and for which motoring pressure is:

$$p_m = p_r \left(\frac{V_r}{V} \right)^k \tag{2.51}$$

A typical value for k is 1.3. The values suggested by Woschni for C_1 and C_2 in the different steps of the cycle are given in Table 2.2. Stone [75] gives particular values for these coefficients in the case of direct injection engines including swirl effects.

Table 2.2 Numerical values for the coefficients C_1 and C_2 suggested by Woschni [71]

Step	C_1	C_2
Intake and exhaust	6.18	0
Compression	2.28	0
Combustion and expansion	2.28	3.24×10^{-3}

4. Hohenberg [79] proposed a modification to Woschni's correlation which was found to underestimate the heat transfer coefficient during compression and to overestimate it during combustion, so the average heat flux in a cycle is overestimated. He proposed a heat transfer coefficient given by:

$$h_{c,g}(t) = \alpha\, V(t)^{-0.06} p(t)^{0.8} T(t)^{-0.4} (v_p + \beta)^{0.8} \tag{2.52}$$

where the constants α and β were fitted by comparing with six representative engines, $\alpha = 130$ and $\beta = 1.4$. Here the instantaneous volume, $V(t)$, is given in m^3, pressure, $p(t)$ is given in bar, temperature, $T(t)$, in Kelvin, and the mean piston speed, v_p, in m/s.

Annand and Ma's model is probably one of the most widely used in the development of quasi-dimensional simulations. Apart from standard isooctane fueled engines, it has been successfully used in simulating engines fueled with alternative mixtures such as hydrogen gasoline [37, 38], and gasoline ethatnol [80]. The empirical correlation by Woschni, apart from pure isooctane, has also been applied to liquified petroleum gas (LPG) powered spark ignition engines [81]. Although the above phenomenological correlations lead to substantially different heat transfer coefficients [63, 73, 82], as we shall see in Sect. 3.3.2, the differences among the resulting performance simulated records for the engine are scarce.

Lounici et al. [74] recently investigated the results of pressure obtained for a two-zone model with different heat transfer models (Annand, Woschni, and Hohenberg) and concluded that for a natural gas engine, Hohenberg's model leads to results closer to experiments. Also recently, Soyhan et al. [83] have systematically analyzed the effects of different heat transfer models for an Homogeneous Charge Compression Ignition (HCCI) engine fueled with gasoline. For direct injection engines, Cho et al. [84] and Fiveland et al. [85] investigated the in-cylinder heat transfer characteristics by precisely measuring the instantaneous surface temperature of the combustion chamber. Two distinct operating modes, homogeneous and stratified, were considered. The suitability of the classic models for heat transfer evaluation was analyzed.

2.3.3 Friction Models

A fraction of the work produced by the gas mixture on the piston does not produce useful work. This nonavailable work is usually defined as *friction work*. This fraction depends on engine design, external load, and some other parameters. The size and design of the coolant system depends to a great extent on the friction forces. Quantitatively, the ratio of friction work ratio to total work may vary from 10 % at full load to 100 % at idle [86]. A good engine design, with reduced frictional losses can lead to better maximum brake torque, reduced specific fuel consumption, more power output, improved durability and reliability, reduced oil consumption, etc. Any prediction of the engine output characteristics requires an estimation of the friction work.

For evaluation purposes, the friction work is defined as the difference between the work delivered to the piston by the gas mixture in the period it is inside the cylinder (compression and expansion strokes) and the useful work to the crankshaft. It has three main components [86]:

1. There is a *pumping component*, W_p, required to pump the fresh mixture inside the cylinder during intake and to expel the residual gases after expansion through the exhaust system.
2. There are many sources of *mechanical friction*, W_{rf}, involving all the moving components of the engine, although many of them are lubricated. These include: friction between the piston rings, piston skirt, and cylinder walls, crankshaft and camshaft bearings, friction in the valve mechanism, gears, etc. It is also known as *rubbing friction work*.
3. Friction arising from the *auxiliary systems*, W_a, such as the fan, water, oil and fuel pumps, the generator, and also the power steering pump and the air conditioning system.

The relative importance of these components depends on the load and speed. The largest one is that associated with the piston and crank assembly, although pumping work may also be considerable. Total friction losses increase with speed. The importance of these terms as well as an up-to-date overview on automotive tribology is given in an elucidating work by Tung and McMillan [87]. From a thermodynamic viewpoint, a recent work by Abu-Nada et al. [88], takes into account variations of the specific heat with temperature of the gas mixture, and presents several results for the dependence of the engine efficiency on speed, oil type, and temperature.

The *total friction work* is defined as the addition of the above components:

$$W_{\text{fric}} = W_p + W_{rf} + W_a \tag{2.53}$$

In order to calculate the power output, it is customary to define the *mean effective pressure*, mep, as the ratio between the total work per cycle, W, and the displaced volume, V_{dt}:

$$\text{mep} = \frac{W}{V_{dt}} \tag{2.54}$$

The power output can thus be calculated from this mean effective pressure as:

$$P = \text{mep}\frac{N}{n_R}V_{dt} \tag{2.55}$$

where n_R is the number of revolutions per cycle ($n_R = 1$ for a two-stroke engine and $n_R = 2$ for a four-stroke engine), and SI units should be used for mep and V_{dt}. Here N is the rotational speed of the crankshaft in rev/s. It is possible to define a mean effective pressure for each of the friction work components, so the *total mean effective pressure*, tfmep, is given by:

$$\text{tfmep} = \text{pmep} + \text{rfmep} + \text{amep} \qquad (2.56)$$

All these quantities are positive, except for pmep, when the pressure inside the cylinder exceeds the external pressure.

In order to numerically evaluate the friction force, F_{fric}, the standard approximation starts by considering that it is proportional to the piston speed, $F_{\text{fric}} = \mu \dot{x}$, where μ is an *effective friction parameter*, around 10 kg/s [23]. Usually, the piston speed is substituted by its mean value. This simple approximation only considers friction between surfaces. The terms associated with friction between surfaces depend on the lubricating film, specifically on its thickness and viscosity. Globally they depend linearly on N, while the terms associated with pumping stem from fluid flow through restrictions and the associated turbulences, and are quadratic functions of N. There are many empirical correlations available in the literature for W_{fric} or tfmep, depending on the rotational speed, such as:

$$W_{\text{fric}} \text{ (or tfmep)} = C_1 + C_2 N + C_3 N^2 \qquad (2.57)$$

However, some friction components depend on the mean piston speed rather than the rotational speed. We summarize some usual phenomenological correlations of this kind from the literature:

1. Barnes-Moss [89, 90] fitted several experimental data from different four-stroke SI engines with displacements from 845 to 2000 cm³. The proposed correlation for tfmep in bar has the following coefficients:

$$
\begin{aligned}
C_1 &= 0.97 \\
C_2 &= 0.15 \times 10^{-3} \\
C_3 &= 0.05 \times 10^{-6}
\end{aligned}
\qquad (2.58)
$$

Considering $N = 60\omega/(2\pi)$, where ω is the rotational speed (SI units) and tfmep $= W_{\text{fric}}/(x\, A_{\text{piston}})$ and $W_{\text{fric}} = F_{\text{fric}}x$, the friction force, F_{fric}, for the Barnes-Moss' correlation can be expressed in SI units as,

$$F_{\text{fric}} = A_{\text{piston}}\left(0.97 \cdot 10^5 + 143.2394488\,\omega + 0.4559453\,\omega^2\right) \qquad (2.59)$$

2. Among the correlations that depend on the mean piston speed, an old but interesting one is that proposed by Chen and Flynn [75, 91] for compression ignition engines[18]:

$$\text{tfmep} = 0.137 + \frac{p_{\max}}{200} + 0.162\, v_P \tag{2.60}$$

where $p_{\max}$ is the maximum pressure in bar, v_P is the mean piston speed (m/s), and tfmep results in bar.

3. Winterbone [92] proposed a similar correlation but by introducing the rotational speed in terms of the mean piston speed:

$$\text{tfmep} = 0.061 + \frac{p_{\max}}{60} + 0.294\, \frac{N}{1000} \tag{2.61}$$

The friction forces are incorporated into the simulations in two alternative ways:

(i) In the case that the external torque, M_{ext}, is a fixed parameter, the mechanical differential Eq. (2.6) has to be solved in conjunction with the thermodynamic equations for each stroke. Friction losses appear directly in Eq. (2.6) as a force against that exerted by the gas. Power output is evaluated through the equation,

$$P = \omega\, M_{\text{ext}} \tag{2.62}$$

(ii) When simulating an engine at a fixed rotational speed, the mechanical Eq. (2.6) is not involved in the dynamic evolution of the system. Under these circumstances, to evaluate the net power output, friction work losses, W_{fric}, and the work performed by the gas mixture on the piston, W_{gas}, have to be independently calculated from the general definition of work. This can be done from the following equations for a four-stroke engine:

$$|W_{\text{fric}}| = \frac{1}{A_{\text{piston}}} \int_0^{4\pi} |F_{\text{fric}}| \left| \frac{dV}{d\varphi} \right| d\varphi \tag{2.63}$$

$$|W_{\text{gas}}| = \left| \int_0^{4\pi} p \left(\frac{dV}{d\varphi} \right) d\varphi \right| \tag{2.64}$$

where $\dfrac{dV}{d\varphi}$ is obtained from Eq. (2.11). Thus, the net work output in a cycle, W, is given by:

---

[18] As discussed by Heywood [90], friction work for compression ignition engines and spark ignition engines is different. Pumping work for SI engines is larger than that for similar CI engines. Piston-crank assembly friction losses are also different. So the use of the correlation for Chen and Flynn [75, 91] for SI engines in order to get accurate performance results should be carefully examined.

$$W = |W_{gas}| - |W_{fric}|$$ (2.65)

and the power output, P, becomes:

$$P = \frac{\omega}{4\pi} W$$ (2.66)

2.3.4 Working Fluid Properties and Chemical Reactions

Other basic ingredients necessary to solve the governing differential equations for each of the strokes (Sect. 2.2) are the thermodynamic properties of the elements forming the working fluid. During intake and compression, the working fluid is composed of fuel, air, and residual gases.[19] Although the fuel can be liquid and vapor in the intake, it is usually taken as vapor. During expansion and exhaust, the working fluid is a gaseous mixture of combustion products.

In this section, we describe how to manage the variables that directly appear in the thermodynamic differential equations under the ideal gas hypothesis. The *ideal gas law* is written as:

$$pv = RT$$ (2.67)

where v is the specific volume and R the *gas constant for each specie*, $R = \tilde{R}/M$. The *universal gas constant* is denoted by $\tilde{R}$, and M is the molecular weight of the gas. The internal energy, u, of an ideal gas is a function of temperature only, $u = u(T)$, and so is enthalpy, $h = u + pv = h(T)$. The specific heats at constant pressure, c_p, and at constant volume, c_v, are respectively, defined as:

$$c_v = \left(\frac{\partial u}{\partial T}\right)_v = \frac{du}{dT}$$

$$c_p = \left(\frac{\partial h}{\partial T}\right)_p = \frac{dh}{dT}$$ (2.68)

The internal energy and enthalpy at any given temperature are obtained by integrating these equations from a *reference state*, T_{ref}.[20]

$$u_{s,i}(T) = \int_{T_{ref}}^{T} c_{v,i}(\xi)d\xi$$

$$h_{s,i}(T) = \int_{T_{ref}}^{T} c_{p,i}(\xi)d\xi$$ (2.69)

---

[19] Some engines incorporate recycled exhaust as a way to control NO_x emissions. They are usually called EGR (exhaust gas recirculation) engines.

[20] Unless explicitly mentioned, we shall take, $T_{ref} = 298.15\,K$.

where the subscript s, i means that these are *sensible* terms for component i in a gas mixture. They are associated with temperature changes and not with energy balances in chemical reactions. When chemical reactions are taken into consideration it is necessary to add a *formation energy*, $\Delta u^\circ_{f,i}$, to the internal energy and a *formation enthalpy*, $\Delta h^\circ_{f,i}$, to enthalpy:

$$u_i(T) = \Delta u^\circ_{f,i} + \int_{T_{\text{ref}}}^T c_{v,i}(\xi)d\xi$$

$$h_i(T) = \Delta h^\circ_{f,i} + \int_{T_{\text{ref}}}^T c_{p,i}(\xi)d\xi \qquad (2.70)$$

From the thermophysical functions of all the components in a gas mixture, it is possible to obtain the properties of a mixture, through the consideration of the mole or mass fractions. For instance, for specific functions and mass fractions, x_i:

$$u = \sum_i x_i\, u_i$$

$$h = \sum_i x_i\, h_i \qquad (2.71)$$

The molecular weight of a mixture of ideal gases, M, can be calculated as:

$$M = \frac{1}{N}\sum_i n_i\, M_i = \sum_i \tilde{x}_i\, M_i \qquad (2.72)$$

where N is the total number of moles, n_i the number of moles of component i, M_i the molecular weight of that component, and $\tilde{x}_i$ its mole fraction.

In principle, it is possible to proceed in an analytical manner by considering that the specific heats are independent of temperature, but this is unrealistic for spark ignition engines, where the working temperature intervals for the unburned or burned gas mixture vary over a wide range (400 to 900 K for the unburned mixture and from 1200 to 2800 K for the burned gases [93]).[21]

There are several references [94, 95], on-line resources [96, 97], and commercial programs [98] that are available to obtain with good accuracy many of the thermophysical or thermochemical properties of almost any chemical specie, particularly in a vapor state. As an example, the *specific heat at constant pressure* of any specie, $c_{p,i}$, considered as an ideal gas can be approximated by a seven parameter polynomial:

$$\frac{c_{p,i}(T)}{R} = a_{i1}T^{-2} + a_{i2}T^{-1} + a_{i3} + a_{i4}T + a_{i5}T^2 + a_{i6}T^3 + a_{i7}T^4 \qquad (2.73)$$

---

[21] Pressure differences are also substantial, from approximately atmospheric pressure at intake to 35 atm at its highest value.

The fitting coefficients, a_{ij}, can be found in [97]. The *specific heat at constant volume* for an ideal gas is obtained from *Mayer's equation*:

$$\frac{c_{v,i}(T)}{R} = \frac{c_{p,i}(T)}{R} - 1 \tag{2.74}$$

In the next sections, we deal with the calculation of enthalpy of the working fluid in a spark ignition engine by distinguishing between the unburned and burned gas mixture.

2.3.4.1 Unburned Gas Mixture

During intake and compression, the working fluid is composed not only of air (a) and fuel (f), but also of residual gases (r), so the enthalpy of the unburned gases per unit mass is given by[22]:

$$h_u = \frac{x_f}{M_f}\tilde{h}_f + \frac{x_a}{M_a}\tilde{h}_a + \frac{x_r}{M_r}\tilde{h}_r \tag{2.75}$$

where M_i is the molecular weight of each component, $\tilde{h}_i$ the corresponding molar enthalpy, and x_i the mass fraction. The mass fractions can be expressed in terms of the fuel–air ratio, ϕ, and the ratio between fuel and air masses under stoichiometric conditions, r_q. It follows that:

$$x_f = \frac{\phi r_q m_a}{m_a \left(1 + \phi r_q\right) + m_r}$$
$$x_a = \frac{m_a}{m_a \left(1 + \phi r_q\right) + m_r}$$
$$x_r = \frac{m_r}{m_a \left(1 + \phi r_q\right) + m_r} \tag{2.76}$$

Here x_r represents the mass fraction of residual gases, *i.e.*, the ratio between the mass of residual gases, m_r, and the total mass inside the cylinder, m: $x_r = m_r/m$. In other words, the total mass inside the cylinder, m, is the inducted mass per cycle, plus the residual mass arising from the previous cycle, m_r.

For most calculations, it is sufficient to take the air as dry air formed by a standard mixture with 21 % mole fraction of oxygen and 79 % of inert gases considered as nitrogen. Actually, there is also a minor fraction of dust and water vapor, depending

---

[22] In some modern engines, a fraction of recycled exhaust gases are injected into the fresh mixture with the objective of a greater dilution and thus control NO_x emissions. To quantify the percent of exhaust gas recycled an additional parameter is introduced: $EGR(\%) = (m_{EGR}/m_i) \times 100$ where m_{EGR} is the mass of exhaust gas recycled. In this case, the residual gas mass fraction in the fresh mixture is: $x_r = (m_{EGR} + m_r)/m$.

Mole fraction (%)	Gas
20.99	Oxygen
78.03	Nitrogen
0.94	Argon, neon, helium, krypton
0.03	Carbon dioxide
0.01	Hydrogen

Table 2.3 Dry air composition expressed as mole fraction

on the atmospheric conditions. The largest water vapor concentration at 21°C and reference pressure is 2.6 % (saturation point). Table 2.3 lists the mole fraction of the principle constituents of dry air. From the data in the table, it is obtained that for each mole of oxygen in air there are 3.717 moles of nitrogen.

With these hypotheses, the molecular weight of dry air is usually taken as $M_a = 28.967$ g/mol. It is also customary to refer to all the inert gases that constitute dry air as *nitrogen*, and to associate an *apparent nitrogen* molecular weight calculated by adding all the weights of the inert gases and dividing by their number of moles, $M_{a,N} = 28.161$ g/mol. The density of dry air, taken as an ideal gas, at $p = 1$ atm and $T = 298$ K, with that value of M_a is $\rho_a = 1.184$ kg/m^3.

2.3.4.2 Burned Gas Mixture and Chemical Reactions

The combustion event in any internal combustion engine is always a fast exothermic gas-phase reaction. The flame front is the reaction zone. Its spatial evolution is a complex process consisting of the chemical reaction, fluid flow, and the transport processes associated with mass diffusion and heat transfer. One of the main elements determining the flame characteristics is the way the reactants enter the reaction zone. If the fuel and the oxidizer are uniformly mixed together before entering the reaction zone, the flame is called *premixed*. On the other hand, if reactants are not previously premixed and they mix in the same region reaction occurs, the flame is classified as a *diffusion* flame. In conventional spark-ignition engines, the flame is premixed and propagates through a gaseous mixture (eventually including combustion products from previous cycles), while in compression ignition engines combustion is characterized by a diffusion flame.[23] Another difference arises from the fact that the fuel–air mixture in spark ignition engines is in the gaseous phase, but in compression ignition engines the fuel enters the cylinder as a liquid. However, a common feature of both combustion processes is that they are unsteady and turbulent.

If we restrict our attention to spark ignition engines, the most important characteristics of the fuel with respect to engine performance, as cited by Taylor [99],

---

[23] This key difference between combustion in both types of engines is more fundamental than the fact that for theoretical purposes a spark ignition engine is associated to an Otto cycle and a compression ignition engine to a Diesel cycle. In reality, combustion never occurs at constant volume in a spark ignition engine, nor does it occur under isobaric conditions in a compression ignition engine.

are: volatility, detonation and pre-ignition characteristics, heat of combustion per unit mass and volume, heat of evaporation, chemical stability, neutrality and cleanliness, and safety. The most widely used fuels are blends of several hydrocarbon compounds obtained by refining petroleum or crude oil. They contain mostly carbon and hydrogen, although there can be traces of several other chemical species. Single hydrocarbon compounds such as methane, propane, and isooctane are the most used. The isooctane is normally taken as the reference fuel.[24] Other fuels of current interest are alcohols and their blends with hydrocarbons, and gaseous fuels such as natural gas or LPG. The basic thermophysical properties of these fuels can be found in the literature or through online resources.

Almost any combustion reaction, involving typical hydrocarbons, alternative fuels, or their blends can be analyzed by means of quasi-dimensional simulations. In the development of the computer codes, there are at least two procedures to incorporate the chemistry of combustion reactions: (a) using an external computer program [96, 97, 100–102] or (b) adding a subroutine in the simulation code itself [103, 104]. In Appendices E and F, we provide some examples of these procedures for pure hydrocarbons and alcohols.

2.4 Assembly of Submodels and Basic Structure of the Simulation

In Fig. 2.7, we present the structure of a generic simulation. The basic elements of a representative simulation are included, but an advanced simulation incorporating more physical or chemical ingredients should have additional submodels linked to the basic elements. We show the connections among the submodels and the skeleton of the simulation (organized around the thermodynamic cycle stages) by arrows. Possibly, a particular simulation could contain more connections. Submodels marked in green colors in this figure contain the hypothesis, analysis, description, and basic parameters of the corresponding phenomena. In some cases, they require specific computations that could be considered as separated parts. These are shown in the figure in rectangles. Precisely, the submodel structure allows one to modify, improve or adapt the simulations to specific requirements without making extensive changes to the whole program. Additional or improved submodels can be incorporated with ease of interchangeability, without altering the overall structure.

The skeleton of the simulations is built around the resolution of the coupled thermodynamic and mechanical differential equations in sequential order, beginning from the induction of the mixture to the exhaust of the residuals shown in blue.

---

[24] The *octane number* of a fuel is a measure of its anti-knock properties. It is defined a 0–100 scale assigning 0 to n-heptane (a fuel predisposed to knock) and 100 to isooctane because of its anti-knock resistance. For instance, a 95 octane fuel has anti-knock performance equivalent to a mixture composed of 95 % isooctane and 5 % n-heptane by volume.

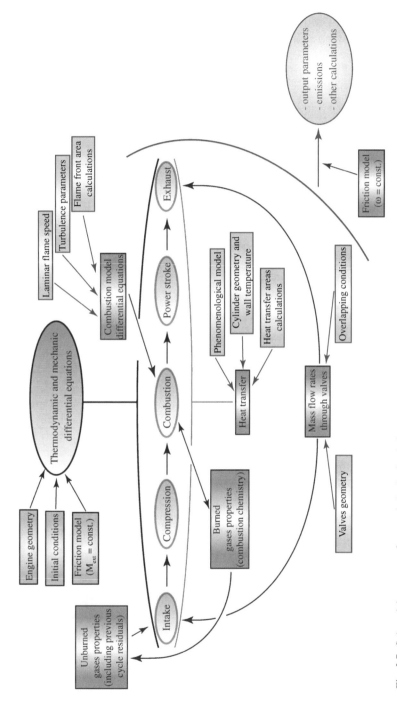

Fig. 2.7 Submodel structure of a generic basic simulation

Because of its key importance, combustion has been considered in the scheme as a separate step, between compression and expansion. As discussed in Sect. 2.3.1, it can be described either phenomenologically or within the quasi-dimensional scheme. In the phenomenological description, a specific differential equation to follow the evolution of the burned gas fraction with time is not required. The parameters associated with the model are some constant coefficients and the combustion duration. Nevertheless, in the quasi-dimensional scheme, another pair of coupled differential equations have to be solved simultaneously with the thermodynamic and mechanical ones. In this scheme, at least correlations for the laminar flame speed, turbulence parameters, and the explicit calculation of the evolution of the flame front inside the cylinder with time are required.

Furthermore, it is necessary to solve the chemistry of combustion. In the simplest approach, the combustion of a lean fuel–air mixture without residuals from the previous cycles is to be solved, and in a more realistic approximation, complex fuel–air residuals reactions with a detailed analysis of emissions (NO_x, HC, etc.) are to be included. In the case residuals are added to the fresh mixture, there will be a chemical memory effect between one cycle and the next. This effect will be important, for instance, in order to reproduce the experimentally observed cycle-to-cycle variations in the quantities such as power output or heat release.

A heat transfer model directly affects all the strokes of the cycle, and a specific calculation of the cylinder walls areas wetted by the burned or unburned gases is essential, due to the important temperature differences. This submodel requires the knowledge of several phenomenological parameters depending on the chosen heat transfer coefficients, the geometry of the cylinder, and a mean cylinder internal wall temperature.

A mass flow rate model is needed during induction and exhaust. In the simplest approximation, as shown in Sect. 2.2.1 and Appendix B, the equations for the quasi-steady flow of a compressible ideal gas through a restriction are sufficient. This first-level approximation can be refined to incorporate turbulence or effects like swirl. Another optional computation related to intake and exhaust is the consideration of overlap, the time span during which both valves are open. Supplementary work is required to detect the sign of the flow rates and the composition of the mixture during the overlap interval. Vandevoorde et al. [105] presents a comparison of different numerical algorithms used in commercial codes for the calculation of the one-dimensional unsteady flow in the pipes of the inlet and exhaust systems of internal combustion engines.

The composition of the mixture and its thermodynamic properties are basic ingredients of any stroke. After selecting the type of fuel and the fuel–air ratio in the fresh mixture, the next step is to consider its thermodynamic properties as temperature and/or pressure-dependent functions. This is usually done in the framework of the ideal gas approximation. The composition of combustion products has to be stored from the previous cycle, which changes the evaluation of the thermodynamic properties for intake and compression.

The friction model plays an important role in the simulation. If the rotational speed is constant, it is incorporated directly in the evaluation of the output parameters of the engine, after solving its dynamics in a cycle (Sect. 2.3.3). Nevertheless, if the external torque is considered as a fixed input parameter, the mechanical Eq. 2.6 has to incorporate a model for the friction losses and then solved together with the thermodynamic equations. The net cycle output work is the difference between the work exerted by the gas on the piston and the friction work. A phenomenological model is normally considered to incorporating several friction mechanisms, but at least a mechanical friction arising from all the moving components of the engine is essential.

With the basic structure and the complementary submodels, it is possible to estimate parameters such as power output, efficiencies, some emissions, and many other parameters or functions. Some other submodels may be added to the basic program structure in order to obtain specific performance data. In Chap. 3, several quantities which can be obtained from quasi-dimensional simulations will be described.

2.5 Numerical Solution of Differential Equations and Submodels

The set of first-order coupled differential equations required to follow the evolution of the thermodynamic variables of the gas mixture in the cylinder (Sect. 2.2) is usually solved with standard algorithms for the solution of ordinary differential equations [106]. It is not easy to find in the literature details about the numerical methods employed, but this is basically because the solution of the equations is not especially problematic. There are references in the literature using fourth-order Runge-Kutta [74, 107] or *predictor-corrector* [7] methods. Usually, the equations are solved taking as independent variable the crankshaft angle with a step of 1° or smaller for particular calculations. With respect to computing time requirements, quasi-dimensional simulations are not very demanding. In a modern PC, hundred of cycles are solved in a time scale of minutes with a crank step of 1°. Pariotis et al. compare in [108] the computation times required by three different models: single-zone, quasi-dimensional, and CFD models.

References

1. J. Heywood, *Internal Combustion Engine Fundamentals* (McGraw-Hill, 1988), Chap. 2, pp. 42–45
2. R. Stone, *Introduction to Internal Combustion Engines* (Macmillan Press LTD., 1999), Chap. 4, pp. 142–155
3. C.R. Ferguson, *Internal Combustion Engines, Applied Thermosciences* (Wiley, 1986), Chap. 1, pp. 1–24
4. J. Heywood, *Internal Combustion Engine Fundamentals* (McGraw-Hill, 1988), Chap. 6, pp. 206–208
5. R. Stone, *Introduction to Internal Combustion Engines* (Macmillan Press LTD., 1999), Chap. 6, pp. 293–297

6. C. Borgnakke, P. Puzinauskas, Y. Xiao, Spark ignition engine simulation models. Technical Report, Department of Mechanical Engineering and Applied Mechanics. University of Michigan. Report No. UM-MEAM-86-35 (1986)
7. H. Bayraktar, O. Durgun, Energy Sources 25, 439 (2003)
8. J. Heywood, *Internal Combustion Engine Fundamentals* (McGraw-Hill, 1988), Chap. 3, pp. 78–81
9. J. Gatowski, J. Heywood, C. Delaplace, Combust. Flame 56, 71 (1984)
10. J. Tagalian, J. Heywood, Combust. Flame 64, 243 (1986)
11. J. Heywood, *Internal Combustion Engine Fundamentals* (McGraw-Hill, 1988), Chap. 14, pp. 766–768
12. R. Stone, *Introduction to Internal Combustion Engines* (Macmillan Press LTD., 1999), Chap. 3, pp. 50–52
13. C. Ferguson, *Internal Combustion Engines, Applied Thermosciences* (Wiley, 1986)
14. A. Alkidas, Energ. Convers Manage. 48, 2751 (2007)
15. P. Blumberg, J. Kummer, Comb. Sci. Tech. 4, 73 (1971)
16. P. Blumberg, G. Lavoie, R. Tabaczynski, Prog. Energ. Comb. Sci. 5, 123 (1979)
17. M. Tazerout, O. Le Corre, Int. J. Appl. Thermodyn. 3(2), 49 (2000)
18. J. Heywood, J. Higgins, P. Watts, R. Tabaczynski, SAE Paper 790291 (1979)
19. E. Sher, T. Bar-Kohany, Energy 27, 757 (2002)
20. H. Bayraktar, O. Durgun, Energ. Convers Manage. 45, 1419 (2004)
21. S. Sitthiracha, S. Patumsawad, S. Koetniyom, *20th Conference of Mechanical Engineering Network of Thailand* (Nakhon Ratchasima, Thailand, 2006)
22. S. Soylu, J. Gerpen, Energ. Convers Manage. 45, 467 (2004)
23. A. Fischer, K. Hoffmann, J. Non-Equilib, Thermodyn. 29, 9 (2004)
24. J. Keck, *in Proceedings of Nineteenth Symposium (International) on Combustion* (The Combustion Institute, Pittsburgh, 1982), pp. 1451–66
25. G. Beretta, M. Rashidi, J. Keck, Combust. Flame 52, 217 (1983)
26. R.J. Tabaczynski, Prog. Energ. Combust. Sci. 2, 143 (1976)
27. R.J. Tabaczynski, F.H. Trinker, B.A.S. Shannon, Combust. Flame 39, 111 (1980)
28. R. Tabaczynski, C. Ferguson, K. Radhakrishnan, SAE Paper 770647 (1977)
29. H. Bayraktar, Renew. Energ. 32, 758 (2007)
30. H. Bayraktar, O. Durgun, Energ. Convers Manage. 46, 2317 (2005)
31. S.R. Turns, *An Introduction to Combustion, Concepts and Applications* (McGraw Hill, 2012)
32. J. Heywood, in *International Symposium COMODIA 94* (Yokohama, 1994)
33. S. Poulos, J. Heywood, SAE Paper 830334 (1983)
34. J. Heywood, *Internal Combustion Engine Fundamentals* (McGraw-Hill, 1988), Chap. 9, pp. 402–406
35. M. Al-Baghdadi, Turkish J. Eng. Env. Sci. 30, 331 (2006)
36. S. Hosseini, R. Abdolah, A. Khani, in *Proceedings of the World Congress on Engineering Vol II* (2008)
37. M. Al-Baghdadi, H. Al-Janabi, Energ. Convers Manage. 41, 77 (2000)
38. H. Al-Janabi, M. Al-Baghdadi, Int. J. Hydrogen Energy 24, 363 (1999)
39. N. Blizard, J. Keck, SAE Paper 740191 (1974)
40. J. Keck, J. Heywood, G. Noske, SAE Paper 870164 (1987)
41. J. Heywood, *Internal Combustion Engine Fundamentals* (McGraw-Hill, 1988), Chap. 14, pp. 771–773
42. G. Andrews, D. Bradley, S. Lwakabamba, Combust. Flame 24, 285 (1975)
43. S. McAllister, J. Chen, A. Fernandez-Pello, *Fundamentals of Combustion Processes* (Springer, 2011)
44. R. Stone, *Introduction to Internal Combustion Engines* (Macmillan Press LTD., 1999), Chap. 8, pp. 361–369
45. S. Verhelst, C. Sheppard, Energ. Convers Manage. 50, 1326 (2009)
46. D. Bradley, M. Haq, R. Hicks, T. Kitagawa, M. Lawes, C. Sheppard, R. Woolley, Combust. Flame 133, 415 (2003)

47. A. Dahoe, J. Loss Prevention Process Ind. **18**, 152 (2005)
48. J. Heywood, *Internal Combustion Engine Fundamentals* (McGraw-Hill, 1988), Chap. 4, pp. 68–72
49. M. Metghalchi, J. Keck, Combust. Flame **48**, 191 (1982)
50. D. Rhodes, J. Keck, SAE Paper 850047 (1985)
51. L.O. Gülder, in *SAE Paper 841000* (1984)
52. P. Gülder, Combust. Flame **56**, 261 (1984)
53. J. Heywood, *Internal Combustion Engine Fundamentals* (McGraw-Hill, 1988), Chap. 9, pp. 371–375
54. J. Heywood, *Internal Combustion Engine Fundamentals* (McGraw-Hill, 1988), Chap. 9, pp. 450–478
55. R. Stone, *Introduction to Internal Combustion Engines* (Macmillan Press LTD., 1999), Chap. 3, pp. 74–75
56. A. Ibrahim, S. Bari, Fuel **87**, 1824 (2008)
57. R. Jenkin, E. James, W. Malalasekera, SAE Paper 972877 (1997)
58. W. Dai, C. Newman, G. Davis, SAE Paper 961962 (1996)
59. C. Rakopoulos, G. Kosmadakis, A. Dimaratos, E. Pariotis, Appl. Energ. **88**, 111 (2011)
60. K. Robinson, N. Campbell, J. Hawley, D. Tiller, SAE Paper 1999–01-0578 (1999)
61. I. Yoo, K. Simpson, M. Bell, S. Majkowski, SAE Paper 2000–01-0939 (2000)
62. J. Heywood, *Internal Combustion Engine Fundamentals* (McGraw-Hill, 1988), Chap. 12, pp. 668–670
63. R. Stone, *Introduction to Internal Combustion Engines* (Macmillan Press LTD., 1999), Chap. 10, pp. 429–433
64. G. Borman, K. Nishiwaki, Prog. Energy Comb. Sci. **13**, 1 (1987)
65. J. Heywood, *Internal Combustion Engine Fundamentals* (McGraw-Hill, 1988), Chap. 12, pp. 670–673
66. R. Stone, *Introduction to Internal Combustion Engines* (Macmillan Press LTD., 1999), Chap. 12, pp. 483–486
67. S. Khalilarya, M. Javadzadeh, Therm. Sci. **14**, 1013 (2010)
68. O.A. Ozsoysal, Energ. Convers Manage. **47**, 1051 (2006)
69. A. Mohammadi, M. Rashidi, in *Proceeding of the 3rd BSME-ASME International Conference on Thermal Engineering* (2006)
70. A. Noori, M. Rashidi, ASME J. Heat Trans. **129**, 609 (2007)
71. G. Woschni, SAE Paper 670931 (1967)
72. G. Eichelberg, Engineering **148**, 463 (1939)
73. M.I. Karamangil, O. Kaynakli, A. Surmen, Energ. Convers Manage. **47**, 800 (2006)
74. M. Lounici, K. Loubar, M. Balistrou, M. Tazerout, Appl. Therm. Eng. **31**, 319 (2011)
75. R. Stone, *Introduction to Internal Combustion Engines* (Macmillan Press LTD., 1999)
76. W. Annand, Proc. I. Mech. E. **177**, 973 (1963)
77. W. Annand, T. Ma, Proc. I. Mech. E. **72**, 976 (1985)
78. N. Watson, J.M.S., *Turbocharging the internal combustion engine* (Macmillan Press LTD., 1982)
79. G. Hohenberg, SAE Paper 790825 (1979)
80. H. Bayraktar, Renew. Energ. **30**, 1733 (2005)
81. H. Özcan, J. Yamin, Energ. Convers Manage. **49**, 1193 (2008)
82. J. Heywood, *Internal Combustion Engine Fundamentals* (McGraw-Hill, 1988), Chap. 12, pp. 694–696
83. H. Soyhan, H. Yasar, H. Walmsley, B. Head, G. Kalghatgi, C. Sorusbay, Appl. Therm. Eng. **29**, 541 (2009)
84. K.W. Cho, D. Assanis, Z. Filipi, G. Szekely, P. Najt, R. Rask, Proc. Inst. Mech. Eng. Part D-J. Automob. Eng. **222**, 2219 (2008)
85. S. Fiveland, D. Assanis, SAE Paper 2002–01-1757 (2002)
86. J. Heywood, *Internal Combustion Engine Fundamentals* (McGraw-Hill, 1988), Chap. 13, pp. 712–713

87. S. Tung, M. McMillan, Tribol Int. **37**, 517 (2004)
88. E. Abu-Nada, I. Al-Hinti, A. Al-Sharki, B. Akash, J. Eng. Gas Turbines Power **130**, 022802 (2008)
89. H.W. Barnes-Moss, *A Designer's Viewpoint in Passenger Car Engines, Conference Proceedings* (Institution of Mechanical Engineers, London, 1975), pp. 133–147
90. J. Heywood, *Internal Combustion Engine Fundamentals* (McGraw-Hill, 1988), Chap. 13, pp. 722–724
91. S. Chen, P. Flynn, SAE Paper 650733 (1965)
92. D. Winterbone, *The Thermodynamics and Gas Dynamics of Internal Combustion Engines* (Oxford University Press, 1986), vol. II, Chap. Transient performance
93. J. Heywood, *Internal Combustion Engine Fundamentals* (McGraw-Hill, 1988), Chap. 4, pp. 109–112
94. NIST-JANAF, Thermochemical Tables, 4th edn. Technical Report, National Institute of Standards and Technology. J. Phys. Chem. Ref. Data (1988)
95. B. Poling, J. Prausnitz, J. O'Conell, *The Properties of Gases and Liquids*, 5th edn.(McGraw-Hill, 2001)
96. NIST Chemistry WebBook. http://webbook.nist.gov/chemistry/
97. B.J. McBride, G. Sanford, Computer program for calculation of complex chemical equilibrium compositions and applications. Users Manual 1311, National Aeronautics and Space Administration, NASA (1996). http://www.grc.nasa.gov/WWW/CEAWeb/
98. E. Lemmon, M. Huber, M. McLinden, NIST standard reference database 23: Reference fluid thermodynamic and transport properties-REFPROP, version 9.0. Tech. rep., National Institute of Standards and Technology, Standard Reference Data, Program (2010)
99. C. Taylor, *The Internal-Combustion Engine in Theory and Practice*, vol. II (MIT Press, 1985), Chap. 4, pp. 128–129
100. NIST Program on Chemical Kinetics. http://www.nist.gov/srd/chemkin.cfm
101. CHEMKIN. http://www.reactiondesign.com/products/open/chemkin.html
102. M. Jelezniak. CHEMKED. http://www.chemked.com/
103. C. Olikara, G. Borman, SAE Paper 750468 (1975)
104. C. Ferguson, *Internal Combustion Engines* (Wiley, 1986), Chap. 3, p. 121
105. M. Vandevoorde, J. Vierendeels, R. Sierens, E. Dick, R. Baert, J. Eng. Gas Turbines Power **122**, 533 (2000)
106. W. Press, S. Teukolsky, W. Vetterling, B. Flannery, *Numerical Recipes. The Art of Scientific Computing*, 3rd edn. (Cambridge University Press, 2007)
107. P.L. Curto-Risso, A. Medina, A. Calvo Hernández, J. Appl. Phys. **104**, 094911 (2008)
108. E. Pariotis, G. Kosmadakis, C. Rakopoulos, Energ. Convers Manage. **60**, 45 (2012)

Chapter 3
Validating and Comparing with Experiments and Other Models

3.1 What Can We Measure from Simulations?

3.1.1 Performance Parameters

3.1.1.1 Power and Work Output

Several standard parameters that are commonly used to quantify the power and work output of an engine were introduced in Sect. 2.3.3. Here, we summarize some of these parameters.

- *Power output.* When the external torque is fixed, the power, P, delivered by the engine is the product of the angular speed, ω, and the external torque, M_{ext}:

$$P = \omega\, M_{ext} \qquad (3.1)$$

  This equation was introduced in Sect. 2.3.3. This measure of power output is usually called *brake power*,[1] because it is the usable power delivered by the engine to the load.
  In the case that the angular speed, ω, is fixed, it is not necessary to solve the mechanical differential equation (2.6) in order to reproduce the engine evolution and so, the external torque M_{ext} is not a required parameter. Consequently, power output should be obtained from Eqs. (2.63)–(2.66) which were presented in the section devoted to friction models.
- The work output is obtained from the pressure data for the gas in the cylinder during the operating cycle. The *indicated work per cycle*, W_{gas}, is calculated by direct integration of the cycle curve in the $p - V$ diagram, Eq. (2.64). It is appropriate to define a *gross indicated work per cycle* as the work delivered to the piston

---

[1] In general the term *brake* added to power, work, or efficiency refers to its magnitude measured at the output shaft, while the term *indicated* refers to the value exerted or computed at the *piston*.

A. Medina et al., *Quasi-Dimensional Simulation of Spark Ignition Engines*, DOI: 10.1007/978-1-4471-5289-7_3, © Springer-Verlag London 2014

over the compression and expansion strokes only, overcoming the pumping work between the piston and the gases in the cylinder during inlet and exhaust strokes. The indicated work per cycle, W_{gas}, is related to the net work output, W, and the brake power, P, through Eqs. (2.65) and (2.66). The *mean effective pressure* (mep) and the *total mean effective pressure* (tmep) were also defined in Sect. 2.3.3. The *net indicated mean effective pressure*, $imep_n$, is the indicated work per cycle, W_{gas}, divided by the displaced volume, V_{dt}: $imep_n = W_{gas}/V_{dt}$.

3.1.1.2 Mechanical and Thermal Efficiencies

- The *mechanical efficiency*, η_m, is the ratio of the net work output, $W = |W_{gas}| - |W_{fric}|$, to the indicated work per cycle, W_{gas}. From Eq. (2.65):

$$\eta_m = \frac{W}{|W_{gas}|} = \frac{|W_{gas}| - |W_{fric}|}{|W_{gas}|} \qquad (3.2)$$

- The *volumetric efficiency*, e_v, is a measure of the effectiveness of the induction stroke. It is defined as the mass of air inducted per cylinder in each cycle divided by the mass of air that occupies the swept volume per cylinder at ambient pressure and temperature.
- Strictly from thermodynamics, the *thermal* or *thermodynamic efficiency* is the ratio of the net work produced per cycle to the energy supplied. This is given by $\eta = W/Q_r$, where Q_r is the energy released by the fuel in the combustion process.
- In order to calculate the *heat release during combustion*, δQ_r, we apply the first principle of thermodynamics for open systems, separating the internal energy variations associated with temperature changes, dU, the work output, δW, and the heat transfer from the working fluid to the cylinder walls, δQ_s:

$$\delta Q_r = dU + \delta W + \delta Q_s \qquad (3.3)$$

where internal energy and heat losses include terms from unburned and burned gases: $U = m_u c_{v,u} T_u + m_b c_{v,b} T_b$ and $\delta Q_s = \delta Q_{s,u} + \delta Q_{s,b}$. Net heat release during the whole combustion process is calculated from the integration of heat release during that period, considering as variable the time or the crankshaft angle. A particular model to evaluate heat losses is required (Sect. 2.3.2).

- A slightly different efficiency definition is that called *fuel conversion efficiency*, η_f. This is the ratio of the net work per cycle to the energy that *could be released* by the fuel during combustion. This energy is given by the mass of fuel supplied to the engine per cycle, m_f, times the *heating value* of the fuel, Q_{LHV}.[2] So

---

[2] It is usual to find in the literature several definitions of heating value, although their numerical values only differ by a few percent [1]. We shall take the *lower heating value at constant pressure* when evaluating the fuel conversion efficiency.

$$\eta_f = \frac{W}{m_f Q_{LHV}} = \frac{P}{\dot{m}_f Q_{LHV}} \qquad (3.4)$$

The difference between η and η_f lies on possible incomplete combustion events where a fraction of the chemical energy escapes through exhaust and it is not effectively transferred to the piston. Such difference may be appreciable either for lean mixtures (because combustion is slower and the fuel may be not completely burned before the exhaust valve opens) or for rich mixtures (because there is no enough oxygen to complete combustion).

3.1.2 Fuel Consumption and Emissions

- The fuel consumption in engine tests is normally measured as the fuel mass flow rate, $\dot{m}_f$. But a more appropriate measure is the *specific fuel consumption* that measures how efficiently the engine transforms the chemical energy associated with the fuel to produce work. It is defined as the fuel mass flow rate per unit power output:

$$\text{sfc} = \frac{\dot{m}_f}{P} \qquad (3.5)$$

The power output obtained per unit mass of fuel entering the cylinder in each cycle, PG, can be obtained as: $PG = P/m_f = 1/(t_{\text{cycle}} \text{ sfc})$. These parameters vary from one cycle to another. Representative values can be obtained as averages for a certain number cycles or taking the last cycle as a prototypical one.

- The gaseous emissions in the engine exhaust consist mainly of nitric oxide and nitrogen dioxide (NO_x), carbon monoxide (CO), unburned hydrocarbons (HC), and particulates. Their concentrations are usually measured in percent by volume, but can also be measured in specific terms by normalizing them with the power output. *Specific emissions* are the mass flow rate of pollutant per unit power output. For instance, $s NO_x = \dot{m}_{NO_x}/P$. Similarly for any other pollutant.

3.1.3 Additional Considerations

One of the main advantages of any kind of simulation in the field of internal combustion engines is their flexibility to compute almost any engine variable *inside* a cycle, or as an average over many cycles. Material and time economy reasons make the simulations a basic developmental tool for the state-of-the-art engines. They can be valuable at the design stage of a particular engine, or can be used to predict the behavior of purely theoretical or future engine models. Simulations can aid in the selection of many optimal parameters of the engine, such as design parameters as well as operating parameters. Evidently, these capabilities are very valuable nowadays where engineers and scientists are trying to design efficient engines with low consumption and low emissions.

By looking inside a cycle, it is possible to follow the evolution of the pressure in the cylinder during all the strokes and thus to construct the $p - V$ diagram, or to measure a basic parameter to predict output parameters such as the peak pressure. Other parameters that can be followed as functions of the crank angle or of the time in each cycle are the temperature, mole fraction of burned gas, masses of burned or unburned gases in the cylinder, flow rates during intake or exhaust, evolution of the angular speed, spatially averaged heat fluxes, size of the flame front, and others.

If we can reconstruct the shape of any of those curves in terms of time or the crank angle, then it is possible to obtain all those parameters that are obtained by their integration during the entire cycle. So simulations allow to compute power or work output, mechanical, thermal or volumetric efficiencies, heat release, consumption or emission parameters, and so on. In addition, it is possible to compute the evolution of these parameters with design or configuration variables of the engine, such as fuel–air equivalence ratio, compression ratio, stroke-bore ratio, spark advance, etc. This is directly linked to the optimization capabilities of the simulations, as we shall see in Chap. 4.

Most of the parameters that are computed over an entire cycle vary from one cycle to another due to several physical and chemical mechanisms such as turbulent flows, turbulent combustion, variations in the instantaneous position of the ignition kernel, and others. These variations are usually called cyclic variations or cyclic variability. So several measured output parameters of the engine are averages with more or less deviations from one cycle to the next. We shall see the possibilities that quasi-dimensional simulations offer to reproduce the experimentally observed cycle-to-cycle variability. This will be the objective of Chap. 5.

The rest of this chapter is devoted to describe some numerical results obtained from quasi-dimensional simulations [2, 3], and to compare them with experimental measurements or with equivalent results from other simulation or theoretical techniques. For the purpose of comparing the various approaches, we shall use a reference engine with the geometric and configuration parameters that Keck et al. [4] used for analyzing turbulent flames in spark ignition engines. Those parameters are given in Appendix G.

3.2 Numerical Results from Zero-Dimensional Models

In this section, we present some numerical results obtained within the framework of zero-dimensional models. Heat transfer from the mixture to the cylinder internal walls is assumed to be correlated by Eichelberg's empirical law, and friction is supposed to have linear and quadratic terms with respect to the angular speed with the parameters given by the correlation by Barnes-Moss, Eq. (2.59). The evolution of the mass fraction of burned gases, x_b, is considered as a Wiebe function. The geometrical

and operating parameters of the engine are taken from the experimental engine analyzed by Keck [4]; see Appendix G.

First, we present the evolution of a typical cycle in a $p - V$ diagram (Fig. 3.1)
for a fixed angular velocity,[3] $\omega = 109$ rad/s, or equivalently, $N = 1044$ rpm. During
compression, pressure increases because of the reduction in volume. After ignition
pressure rapidly increases, its slope until the mixture reaches approximately a minimum volume. As a consequence of the fast pressure rise, the mixture exerts a force
on the piston and the gases expand during the power stroke. Finally, the exhaust
valve opens and pressure is reduced. Exhaust and intake are approximately isobaric
processes, because the valve lift is large enough to cause no significant pressure
changes.

A comparison of the pressure variations in the cylinder during the combustion
period with experimental results is depicted in Fig. 3.2. In this figure and in all the
others in this chapter, the spark angle is set at $330°$. The agreement between the
predictions of the zero-dimensional model and the bench engine tests [4] is noteworthy. The evolution of pressure is strongly linked to the spark timing. Advanced
ignition causes a boost of pressure that slows down the piston in its upward motion.
On the other hand, late ignition that moves the piston downward is accompanied by

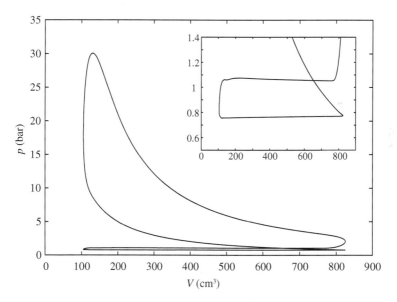

Fig. 3.1 $p - V$ diagram obtained from a zero-dimensional model using the Wiebe function to
represent the evolution of burned gases during combustion. The inset represents a zoom of the
intake and exhaust processes

---

[3] The cycle shown in Fig. 3.1 or in other figures in this chapter is an arbitrary one during the engine
evolution. As it will be detailed in Chap. 5 appreciable oscillations in pressure or other representative
variables can occur from one cycle to the following. The physical origin of this cyclic variability is
quite complex and will be analyzed in that chapter.

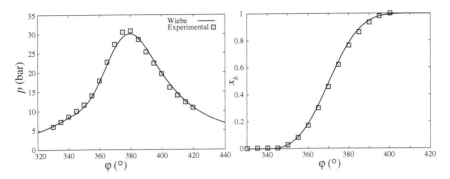

Fig. 3.2 Evolution with the crankshaft angle, φ of the pressure, p, inside the cylinder, and the mole fraction of the burned gases, x_b. The experimental curves [4] and those obtained from the zero-dimensional simulations assuming a Wiebe function for x_b are shown

the expansion of gases, thus decreasing the pressure. In this way, the engine torque is reduced. This is the reason to look for a compromise spark timing. Most of the engine performance parameters can be described in pressure evolution diagrams.

In the case that the external torque is considered as a fixed parameter, the mechanical Eq. (2.6) is solved together with the thermodynamic equations. The engine accelerates from the startup until reaching a stationary regime (see the inset in Fig. 3.3). The left panel in Fig. 3.3 shows that before the stationary regime the speed first decreases until the beginning of combustion when it rapidly increases in such a way that after the four strokes the angular speed shows a net increment with respect to the initial speed in such cycle. Even when the engine operates at a constant speed, the speed is not constant within a cycle, but decreases during compression, then increases rapidly from the ignition event and then exhibits an oscillatory behavior

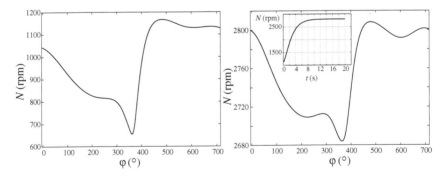

Fig. 3.3 *Left panel* angular velocity of the engine, N, as a function of the crankshaft angle, φ throughout a cycle close to startup when the external torque is a fixed parameter. *Right panel* $N(\varphi)$ for a cycle when the engine reached the stationary regime. The inset shows the evolution of N from startup up to the stationary regime

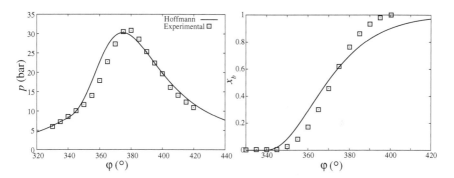

Fig. 3.4 Evolution with the crankshaft angle, φ of the pressure, p, inside the cylinder, and the mole fraction of the burned gases, x_b. The experimental curves and those obtained from the zero-dimensional model by Hoffmann and Fischer [5] are shown

(right panel of Fig. 3.3). But in this case, the initial and final speeds in the cycle are very similar.

An alternative zero-dimensional model without use of the Wiebe function is proposed by Hoffmann and Fischer in [5] and is discussed in Sect. 2.3.1. The evolution with time of the mole fraction of burned gases, $x_b(t)$, is given by Eq. (2.30). The comparison of the evolution of pressure and $x_b(\varphi)$ with the experimental results by Keck are shown in Fig. 3.4. The parameter β_c in Eq. (2.30), was fitted in order to approximately reproduce the maximum experimental pressure. The results show that although the model qualitatively reproduces the experimental results, the agreement between model predictions and experiments is far from satisfactory. The increase of x_b with φ is too rapid (right panel in Fig. 3.4), which leads to an overestimation of pressure from roughly 350° to peak pressure location.

In order to check how the differences in the evolution of pressure during combustion are reflected in the engine output parameters we have compared in Tables 3.1 and 3.2 the predictions of power output, efficiency, and some consumption parameters obtained either from the Wiebe model for combustion or using the evolution of x_b as proposed by Hoffmann and Fischer [5]. For a fixed angular speed (Table 3.1), differences do not exceed 7 % in power output or efficiency. But the differences are

Table 3.1 Performance parameters predicted for two different zero-dimensional models fixed angular speed, $N = 1044$ rpm

	Wiebe	Hoffmann	Diff. (%)
P (kW)	4.563	4.871	6.5
η	0.353	0.373	5.5
imep$_n$ (kPa)	727.656	776.813	6.5
$\dot{m}_f$ (kg/h)	1.042	1.051	0.9
PG (kW/g)	137.210	145.347	5.8

Table 3.2 Output parameters predicted for two different zero-dimensional models and three different external torques

	$M_{ext} = 35\,\text{Nm}$		$M_{ext} = 30\,\text{Nm}$		$M_{ext} = 25\,\text{Nm}$	
	Wiebe	Hoffmann	Wiebe	Hoffmann	Wiebe	Hoffmann
P (kW)	11.192	7.095	12.317	7.113	12.397	6.860
η	0.325	0.300	0.303	0.260	0.280	0.221
$imep_n$ (kPa)	630.994	630.620	540.741	540.382	451.755	450.794
$\dot{m}_f$ (kg/h)	2.737	1.837	3.239	2.150	3.533	2.457
PG (kW/g)	357.597	209.726	428.789	212.729	477.351	208.610
N (rpm)	2953.183	1873.117	3792.281	2191.751	4568.931	2533.706
t_{sim} (s)	20.018	20.033	20.007	20.018	20.006	20.013

t_{sim} represents the simulated time in order to stabilize the engine speed

larger for fixed external torque (Table 3.2). Except for efficiency, where differences are similar to the case $N = $ constant, other differences are in the interval 30–40 %. To a large extent, these differences are associated with the fact that the angular speed required for those external torques is quite different in both models. They are considerably larger from the Wiebe-function model. So all those output parameters related with the stationary angular speed become appreciably different.

3.3 Validating a Quasi-Dimensional Simulation Model

In this section, we present the results obtained from a quasi-dimensional model and compare them with the reference experimental results of Keck [4]. An analysis similar to the one that will be presented here should be done in order to validate a newly developed simulation. In each case, depending on the available experimental results, the comparison between real measurements and simulated predictions could be quantitative or qualitative.

3.3.1 Quasi-Dimensional Model by Keck and Beretta

In this section, we compare the predictions of a quasi-dimensional simulation that model combustion through the entrainment model developed by Keck and Beretta [4, 6] with real engines. Direct quantitative comparisons will be shown in the case experimental results are available, otherwise the comparison will be qualitative, putting emphasis on whether the simulations are capable of reproducing the main physical facts involved in the system evolution. We shall emphasize the physical interpretation of the curves and tables presented in order to give basic guidelines to check the efficacy of the results obtained from other simulations. Next the main hypotheses associated with this kind of simulations are summarized.

- The thermodynamic properties are considered as homogeneous in each control volume.
- During each period in which chemical reactions do not occur, enthalpy changes are only associated with temperature changes.
- Except during combustion, an adiabatic gas mixture is composed of unburned and burned gases.
- Throughout combustion, burned gases are separated from the unburned gases through an adiabatic flame front with negligible volume.
- When all valves are closed, the system is closed, and there are no mass exchanges with the cylinder surroundings.
- Gases are considered as ideal. Their internal energy as well as their enthalpy are functions only of temperature. The gas constants depend only on their compositions.
- The fuel–air ratio, ϕ, is constant. The chemical composition of air and fuel remains unchanged during the engine evolution.
- In the combustion model [4, 6], not all the mass inside the flame front is completely burned, but there exist unburned eddies characterized by a typical size (see details in Sect. 2.3.1).

The evolution of pressure with the crankshaft angle during combustion is shown in Fig. 3.5, together with its experimental counterpart [4]. In the particular results shown, the chemical reaction scheme includes 10 chemical species and dissociation. We see that in all the intervals comparison is remarkably good, not only with respect to the maximum pressure but also its rise and decay. Actually, the predictions of this model for the considered reference engine geometry are considerably better than those of the zero-dimensional models we analyzed in Sect. 3.2 (see Figs. 3.2 and 3.4). So it is apparent that any output parameter directly related to pressure evolution can be reproduced with a fair degree of accuracy within this simulation scheme.

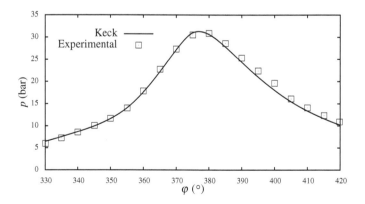

Fig. 3.5 Evolution of the pressure inside the cylinder with the crankshaft angle as predicted by a quasi-dimensional simulation with the model by Keck and Beretta for combustion

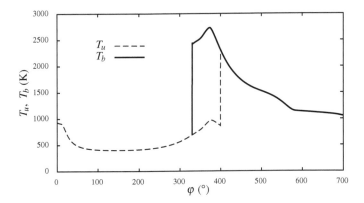

Fig. 3.6 Simulated evolution of the temperatures of the burned, T_b, and unburned gases, T_u

It is also possible to obtain the temperature of the unburned and burned gases during an entire cycle. This is shown in Fig. 3.6. During induction, the temperature of the fresh mixture, T_u, decreases due to induction of the gases from the intake manifold. Once the intake valve closes, the temperature increases as pressure increases, and the volume decreases. Immediately after the ignition event, the temperature of the burned mixture becomes the adiabatic flame temperature. As considered in this kind of simulations, the combustion temperatures are different for the burned and unburned mixtures, corresponding to the two curves depicted in the figure. During the time interval, the volume continues to increase the temperature (T_b) and pressure also increase. After combustion, T_b decreases, and the slope of the curve experiences a change when the exhaust valve opens, because of the expansion of the gas. At the end of combustion, the fraction of the mixture that was not completely burned exchanges heat adiabatically with the burned mixture, so the temperatures become equal.

The evolution of the masses of unburned and burned gases during the whole cycle is elucidating. This is illustrated in Fig. 3.7. The cylinder filling up, the combustion period, and the exhaust process are clearly shown. The evolution of masses during combustion (shaded region in Fig. 3.7) is symmetric, mathematically the time derivatives $\dot{m}_u$ and $\dot{m}_b$ have opposite signs, unburned gases are being consumed resulting in burned gases.

We now consider the mass flow during intake, $\dot{m}_{in}$, see the upper panel of Fig. 3.8. Immediately after the intake valve opens, the flow rate is negative due to the pressure differences from inside the cylinder to the intake manifold. Thereafter, the mass enters the cylinder ($\dot{m}_{in} > 0$) and reaches a maximum when the intake valve lift is maximum and then decreases to approach zero when the crankshaft angle is 180°, and the volume begins to decrease.

The lower panel of Fig. 3.8 illustrates the evolution of the mass flow during exhaust, $\dot{m}_{ex}$. It has an oscillatory behavior due to the changes in the thermodynamic state of the mixture and the speed it reaches. When the exhaust valve opens,

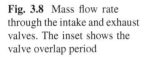

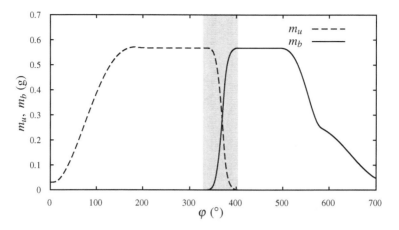

Fig. 3.7 Typical evolution of the masses of the burned, m_b, and unburned gases, m_u, inside the cylinder during a cycle. The shaded region corresponds to combustion

Fig. 3.8 Mass flow rate through the intake and exhaust valves. The inset shows the valve overlap period

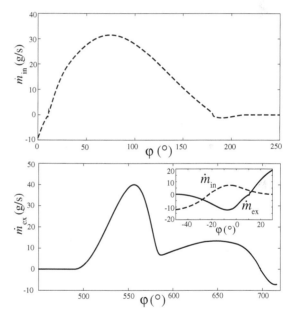

the pressure inside the cylinder is much higher than that at the exhaust manifold, and the gas flow reaches the speed of sound (Mach number over 1). Thereafter, the flow is limited until the speed is lower than the speed of sound (Mach number below 1). From that moment, the flow evolves in a way similar to that during intake.

In the inset of Fig. 3.8, the evolution of the mass flows is shown during the period both valves, intake and exhaust, are open together. This is the so-called overlap period. During this period, the flow depends upon three different pressure values:

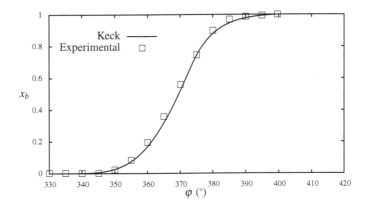

Fig. 3.9 Burned gases mass fraction, x_b, during combustion

the pressure inside the cylinder, p, the pressure in the intake manifold, p_{in}, and the pressure in the exhaust manifold, p_{ex}. Initially, the flow rate through the intake valve is negative because $p > p_{in}$, and also the flow rate is negative at the exhaust valve, $p > p_{ex}$. So the flow goes out of the cylinder through both valves. At the exhaust valve, this happens up to $p = p_{ex}$ and then a change of sign occurs. Similarly, at the intake valve the flow rate changes its sign when $p = p_{in}$. Eventually, if $p_{in} \simeq p_{ex}$, both flow rates change their sign approximately at the same time.

Next we perform another quantitative comparison with experimental results. In Fig. 3.9, the evolution of the burned gas mass fraction, x_b, during the combustion period as obtained from simulations (solid curve) is compared with the experimental results by Keck et al. [4] for the same engine geometry. The quasi-dimensional model reproduces the experimental behavior quite accurately, not only at the beginning and the end of combustion but also during the period when x_b rapidly increases.

From the experimental results, it is also possible to follow the evolution of the flame front radius, R_f, by assuming that the flame front is approximately spherical. Both experimental results and results from quasi-dimensional simulations are shown in Fig. 3.10. Except for the experimental data to the right, the simulations adequately predict not only the evolution but also the numerical values of R_f. The mentioned discrepancy does not affect the predictions for the thermodynamic variables of the system.

The relationship between the flame front area, A_f, and its volume, V_f is depicted in Fig. 3.11. The behavior is as expected from experimental results for a noncentered ignition kernel. The inset in the figure shows the experimental evolution of the nondimensional counterpart of A_f against V_f for several values of the ratio between distance from the cylinder center to the spark plug position, R_c, and the cylinder radius, $B/2$. For intermediate values of this ratio, the curve shows a single maximum point that corresponds to $\alpha = 90°$ (a top view scheme of the evolution of the flame front is shown in the bottom panel of Fig. 3.11). For a centered ignition ($R_c = 0$), A_f monotonically increases with V_f, and at the other extreme, when $R_c/(B/2) \simeq 1$

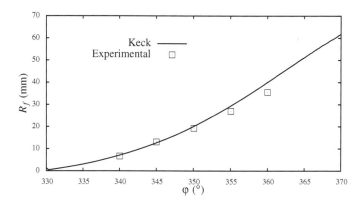

Fig. 3.10 Flame front radius evolution as obtained from the quasi-dimensional model by Keck and Beretta

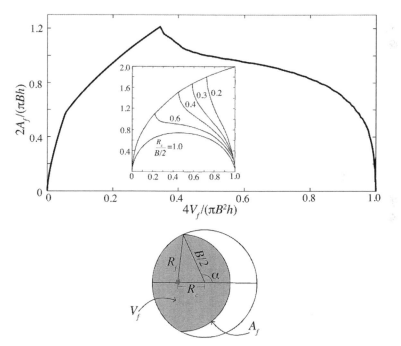

Fig. 3.11 A dimensional representation of the flame front area, A_f, as a function of its volume, V_f, obtained from simulations. The inset corresponds to the experimental results by Keck et al. [4] for several values of the ratio between the spark location and the cylinder radius, $R_c/(B/2)$. Simulation results (*thick solid line*) were obtained for $R_c/(B/2) = 0.39$. The *lower* picture is a *top view* scheme with the main parameters characterizing A_f for a noncentered ignition. R_c represents the spark location with respect to the cylinder head center, B, the cylinder bore, and R_f, the instantaneous flame front radius

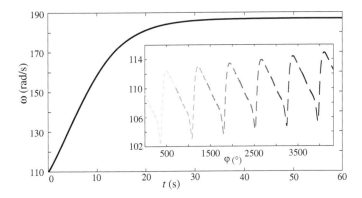

Fig. 3.12 Evolution with time of the angular speed, ω, when the external torque is a fixed parameter obtained from a quasi-dimensional simulation. The development of ω for several engine cycles is also shown as a function of the crankshaft angle, φ

(spark plug position close to cylinder walls) the curve shows a maximum (non-singular) at a nondimensional volume, $4V_f/(\pi B^2 h)$, around 0.75. Results from the quasi-dimensional simulations enable us to reproduce this characteristic evolution of A_f.

If the condition of a constant angular velocity, ω, is substituted by a fixed external torque, M_{ext}, ω evolves from an initial value (usually given by an electric engine starter) to an stationary speed. Although from a global perspective, the evolution seems apparently monotonic (solid curve in Fig. 3.12), in any engine cycle ω oscillates in such a way that for the first period (before reaching the stationary state) there is a net increase of the maximum speed from one cycle to the next (inset in Fig. 3.12), whereas at large times, the peak value of ω remains constant from one cycle to another. It is worth mentioning that the curves shown in Fig. 3.12 were obtained by simulating a single cylinder engine in which the oscillatory evolution of ω for each cycle presents a larger amplitude. For multi-cylinder engines, combustion in the cylinders is synchronized in order to get an output power as uniform as possible. Some performance and consumption parameters for several values of M_{ext} are given in Table 3.3.

3.3.2 Influence of Heat Transfer Models

In this section, we show the influence of the heat transfer model selected to obtain numerical results from the simulations. Annand's correlation, Eq. (2.48), globally predicts a higher heat transfer coefficient (Fig. 3.13) than Eichelberg's, Eq. (2.47), and Woschni's formulas, Eq. (2.49). These differences lead to dissimilar curves for pressure and temperature evolutions.

Table 3.3 Output parameters predicted from the quasi-dimensional model by Keck and Beretta at three different external torques

	$M_{\text{ext}} = 35\,\text{Nm}$	$30\,\text{Nm}$	$25\,\text{Nm}$	$20\,\text{Nm}$
P (kW)	7.336	6.993	6.859	6.196
η	0.297	0.259	0.224	0.186
imep_n (kPa)	618.757	528.691	439.347	355.018
$\dot{m}_f$ (kg/h)	1.975	2.168	2.458	2.668
PG (kW/g)	218.757	212.177	216.506	201.827
N (rpm)	1973.145	2201.319	2598.359	2904.839
t_{sim} (s)	20.032	21.071	20.011	20.032

t_{sim} represents the simulated time in order to reach an stationary regime

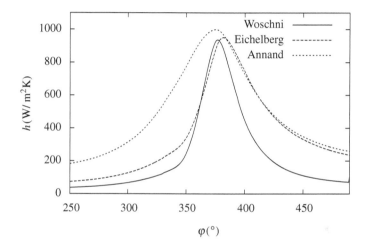

Fig. 3.13 Heat transfer coefficients h, obtained from quasi-dimensional simulations with different empirical correlations

Figure 3.14 shows that the use of Woschni's correlation produces a higher pressure during the entire combustion period. This is due to the fact that heat transfer losses associated with this model are the lowest among the models considered, so the temperature inside the cylinder is higher (see Fig. 3.15). For the particular engine geometry used, Woschni's correlation gives the best approximation to the experimental variations of the pressure in the cycle.

An important effect related to the heat transfer to the cylinder walls is that on residual gases. If a certain correlation predicts higher pressures, during exhaust a larger amount of residual gases will flow out of the chamber. Moreover, their temperature will also be higher. This implies that in the following cycle the combustion chamber will contain less residual gases, thus improving the efficiency.

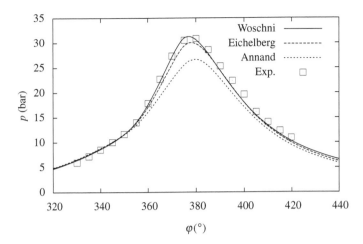

Fig. 3.14 Effect of different heat transfer models s in the evolution of the pressure for a typical simulated cycle. Note particularly, the fair agreement of Woschni's model when comparing with the experimental results

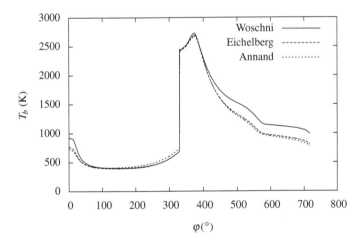

Fig. 3.15 Evolution of the burned gases temperature, T_b as predicted from different heat transfer laws

3.3.3 Results with Alternative Fuels

All the results shown in this chapter until this point were obtained by considering isooctane as fuel. The objective of this section is to briefly present explicit comparisons between simulation and experimental results obtained from alternative fuels. Particularly, we show in Fig. 3.16 results for gasoline-ethanol blends at different concentrations of ethanol by volume and the corresponding experimental values [7]

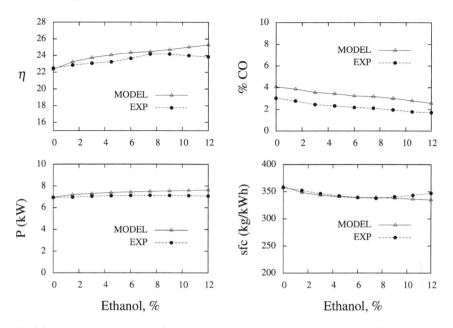

Fig. 3.16 Comparison of the results from simulation with the experimental ones of Bayraktar [7] for gasoline-ethanol blends and different concentrations of ethanol by volume

(details on the chemical reactions involving alternative fuels can be found in Appendices E and F). The addition of ethanol to gasoline raises engine volumetric efficiency and causes leaner operation. As a result, combustion becomes more complete or more stoichiometric, increasing maximum cylinder temperature and pressure. Thus, mean indicated work and mean indicated pressure increase as also do power output and thermal efficiency.

The simulation predicts an engine effective efficiency increasing, although experimental results report a quite flat efficiency maximum at around 8 % ethanol by volume. In any case, the results are in good agreement with other in the literature [7–11]. In an analogous way, experiments show a flat maximum for power about 7.5 % ethanol, but quasi-dimensional simulations predict a monotonic steady increase with ethanol percentage. Differences between model results and experiments are never over 6.5 %.

Carbon monoxide emissions decrease as the volume fraction of ethanol increases due to two facts: more complete combustion and increased combustion temperature as ethanol concentration increases and lower carbon content of ethanol compared with gasoline. The theoretical predictions are in accordance with the evolution with the percentage of ethanol of the experiments, although an overestimation around 25 % is found in all the intervals considered. This is probably due to the fact that the composition is calculated under chemical equilibrium.

With respect to the specific fuel consumption, the addition of ethanol reduces the heating value of gasoline-ethanol blends, so in principle more fuel per mass unit is needed to obtain power outputs similar to those for an engine working with pure gasoline. But the operation with blends makes it leaner, thus improving combustion. For these reasons, the evolution of sfc with the percentage of ethanol shows a small decrease. It is monotonic from simulations and it has a very flat minimum from experiments. In any case, numerical differences between both results are quite small. In summary, the quasi-dimensional simulation is capable to reproduce the numerical values and evolution with the volume percentage of ethanol of the main performance parameters of the engine as well as the presence of CO on exhaust and the specific fuel consumption.

3.4 Finite-Time Thermodynamics Approach

In this section, we present a finite-time thermodynamics (FTT) approach with the aim to reveal another role that simulation models can play in the research and optimization of internal combustion engines: to improve thermodynamic theoretical models in order to make their predictions closer to real engines [5, 12]. Estimations from simulations with respect to engine output parameters will be directly compared with theoretical predictions made from the FTT formulation. A detailed description of the FTT approach was given in Sect. 1.2.

It is instructive to begin by analyzing the well-known reversible air-standard Otto cycle with two isochoric processes (combustion and cooling by means of infinite external heat sources) alternating with two adiabatic processes (compression and expansion). Intake and exhaust processes are assumed to be isobaric and with identical pressures, so they do not influence the calculation of power or efficiency. Figure 3.17 shows the $p - V$ diagram and its comparison with a simulated curve.

Thermodynamically, this idealized cycle is described in terms of two independent variables: the compression ratio $r = V_1/V_3 \geq 1$ and the ratio between the minimum and maximum temperatures, $\tau = T_1/T_3 \leq 1$. In terms of these variables, it is straightforward to obtain the work output, $|W_{rev}| = C_v T_3 \left[(1 - \tau r^{\gamma-1}) - \left(r^{1-\gamma} - \tau \right) \right]$ and a first-law efficiency, $\eta_{rev} = 1 - r^{1-\gamma}$. Clearly, the power output is zero because of the reversible nature of the cycle.

Under maximum work conditions, the efficiency is given by the Curzon–Ahlborn value: $\eta_{CA} = 1 - \sqrt{\tau}$ [13].[4] This is depicted in Fig. 3.18, a pictorial scheme that it is not usually shown in textbooks. The nondimensional work output $(\overline{W})$ has positive values between $r = 1$ and $r = \tau^{1/(1-\gamma)}$, and has a parabolic shape with

---

[4] The publication in 1975 of Curzon–Ahlborn's pioneering work [13, 14] opened the perspective of establishing more realistic theoretical bounds for real energy converters, and give raise to the birth and development of thermodynamic optimization, particularly FTT [15–18]. They showed that a Carnot heat engine coupled to external reservoirs through heat transfers that obey the Fourier law has an efficiency at maximum power given by $\eta_{CA} = 1 - \sqrt{\tau}$. This expression provides a surprisingly good approximation to the observed efficiencies of very different engines [19–22].

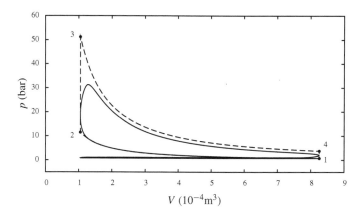

Fig. 3.17 $p - V$ diagram of an ideal Otto cycle (*dashed*) and a simulated (*realistic*) one (*solid*)

Fig. 3.18 Evolution of the thermal efficiency and the dimensional work output, $|\overline{W}| = |W|/C_v T_3$, with the pressure ratio r for the ideal Otto cycle

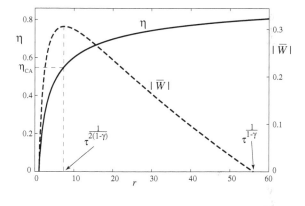

maximum at $r_{max} = \tau^{1/2(1-\gamma)}$. This compression ratio corresponds to an efficiency $\eta(r_{max}) = 1 - \tau^{1/2} = \eta_{CA}$ that is the Curzon–Ahlborn efficiency [13].

FTT irreversible models improve the reversible version by adding three main types of energy losses: friction work associated with the piston dynamics; heat transfer from the working fluid to the surroundings through the cylinder walls; and those associated with the working fluid and which account for fluid friction, viscosity, turbulence, and so on. On this basis, the work performed by the engine can be expressed as:

$$|W| = |W_{rev}| - (|W_{rev}| - |W_I|) - |W_{fric}| - |W_Q| \qquad (3.6)$$

where $|W_I|$ is the work output including internal irreversibilities, so $|W_{int}| = |W_{rev}| - |W_I|$ accounts for energy losses due to internal irreversibilities. $|W_{fric}|$ represents the energy loss due to friction, and $|W_Q|$ the energy loss due to heat transfer across the cylinder walls. So

$$|W| = |W_{rev}| - |W_\ell| \qquad (3.7)$$

where $|W_\ell|$ represents the net work losses, $|W_\ell| = |W_{\text{int}}| + |W_{\text{fric}}| + |W_Q|$.

Besides these components of work losses, the FTT-model may include two specific characteristics associated with the combustion process [23, 24]: changes in the heat capacities before and after combustion, and a temperature T_3 considered as the adiabatic flame temperature. Next, we discuss these characteristics and show how to evaluate the energy losses.

3.4.1 Internal Irreversibilities

A common way to include internal irreversibilities in FTT models is based on the idea by Özkaynak et al. [25], and Chen [26], for Carnot-like cycles by using the Clausius inequality. If $|Q_H|$ ($|Q_C|$) denotes the heat absorbed (released) for the working fluid from the hot reservoir (to the cold reservoir) at temperatures T_H (T_C) we can write the Clausius inequality as $I_R \dfrac{|Q_H|}{T_H} - \dfrac{|Q_C|}{T_C} = 0$, where the parameter $I_R > 1$. Thus, the efficiency is given by:

$$\eta = 1 - I_R \frac{T_C}{T_H} = 1 - I_R \left(\frac{|Q_C|}{|Q_H|} \right)_{\text{rev}} = \frac{|W_I|}{|Q_H|_{\text{rev}}} \tag{3.8}$$

This enables us to obtain $|W_I|$ as $|W_I| = |Q_H|_{\text{rev}} - I_R |Q_C|_{\text{rev}}$. Although in Otto models heat transfers do not occur under isothermal conditions, FTT-models assume that it is so, and $|Q_H|_{\text{rev}} = |Q_{23}|$ and $|Q_C|_{\text{rev}} = |Q_{41}|$ (see Fig. 3.17). Then, we obtain,

$$|W_I| = T_3 \left[\overline{C}_{v,23} \left(1 - \tau r^{\overline{\gamma}_{u,12}-1} \right) - I_R \overline{C}_{v,41} \left(r^{\overline{\gamma}_{b,34}-1} - \tau \right) \right] \tag{3.9}$$

where the heat capacities and adiabatic coefficients are, in general, calculated as averages over the considered temperature interval. For instance, taking into account the chemical composition of the gas mixture in the cylinder before (u) and after (b) combustion,

$$\overline{C}_{v,23} = \frac{1}{2} \left[C_{v,u}(T_2) + C_{v,b}(T_3) \right]; \quad \overline{C}_{v,41} = \frac{1}{2} \left[C_{v,b}(T_4) + C_{v,u}(T_1) \right] \tag{3.10}$$

and

$$\overline{\gamma}_{u,12} = \frac{\overline{C}_{p,12}}{\overline{C}_{v,12}} = \frac{C_{p,u}(T_1) + C_{p,u}(T_2)}{C_{v,u}(T_1) + C_{v,u}(T_2)}; \quad \overline{\gamma}_{b,34} = \frac{\overline{C}_{p,34}}{\overline{C}_{v,34}} = \frac{C_{p,b}(T_4) + C_{p,b}(T_3)}{C_{v,b}(T_4) + C_{v,b}(T_3)} \tag{3.11}$$

3.4.2 Friction Losses

In FTT models, the friction work associated with the piston motion inside the cylinder is usually calculated on the basis of a mean piston speed which is obtained from the knowledge of the extreme piston positions, and the duration of the power stroke considered as a fraction of the whole duration of the cycle [23, 27, 28]. Here, we consider a more realistic situation with a frictional force, $|F_{\text{fric}}| = \mu|\dot{x}|$, proportional to the instantaneous velocity $|\dot{x}|$, with a constant friction coefficient, μ. The velocity can be written in terms of the crank radius, a, the crank angle, φ, and the angular velocity, ω, through the function $\xi_1(\varphi)$ defined in Eq. (2.7):

$$|\dot{x}| = a\omega\xi_1(\varphi) \tag{3.12}$$

Thus, the total friction work is calculated from Eq. (2.63). Considering Eq. (2.11),

$$\frac{dV}{d\varphi} = \frac{V_0}{2}(r-1)\xi_1(\varphi) \tag{3.13}$$

we finally get:

$$|W_{\text{fric}}| = \frac{a\mu\omega V_0(r-1)}{2A_{\text{piston}}} \int_0^{4\pi} \xi_1^2(\varphi)\,d\varphi \tag{3.14}$$

3.4.3 Heat Transfer Losses

It is assumed that the heat transfer from the working fluid to the surroundings has a convective nature and that it is only relevant during the power stroke, where the piston position and fluid temperature are considered as average values, $\bar{x}_{34}$ and $\overline{T}_{34}$. Then, assuming a cylindrical combustion chamber, the average heat transfer, $\overline{Q}_l$, can be written as [29–31]:

$$\overline{Q}_l \simeq \pi h B \left(\frac{B}{2} + \bar{x}_{34}\right)(\overline{T}_{34} - T_w)\,t_{34} \tag{3.15}$$

where h is the heat transfer coefficient, T_w the wall temperature, B the cylinder bore, and t_{34} the duration of the power stroke.

For the evaluation of the energy losses associated with heat transfer, Mozurkewich and Berry [32] assume that $W_Q = \varepsilon\overline{Q}_l$ where ε is a phenomenological constant ($\simeq 0.1$), and that t_{34} is $t/4$ times the whole duration of the cycle, t (that can be obtained from the angular speed, ω). So $|W_Q|$ is given by,

$$|W_Q| = \frac{\pi \varepsilon h B t\, T_3}{16} \left[B + \frac{V_0}{A_{\text{piston}}} (1+r) \right] \left(1 + r^{1-\gamma} - 2\frac{T_w}{T_3} \right) \qquad (3.16)$$

After having analytical expressions for W_I, W_{fric}, and W_Q, it is possible to obtain the net work output, $|W|$, from Eq. (3.7), the power output as $P = |W|/t$ and the efficiency as $\eta = |W|/|Q_{23}|$.[5]

3.4.4 Numerical Application

In most FTT models both the temperature after combustion, T_3, and the inlet temperature, T_1, are considered as input parameters [29] or estimated by assuming particular characteristics for heating and cooling branches [27, 33, 34]. Another option is to consider T_3 as the *adiabatic flame temperature*. It is calculated by supposing that all internal energy, chemical and sensible, in reactants $U_R(T_2)$, is transferred to products $U_P(T_3)$ [23]. Then, T_3 is obtained from $U_R(T_2) = U_P(T_3)$. In turn, T_2 is obtained from the adiabatic equation for the path $1 - 2$, $T_2 = T_1 r^{\overline{\gamma}_{12}-1}$, so it is possible to consider just T_1 as an input parameter to obtain the combustion temperature, T_3.

In order to evaluate the results from the analytical FTT formulation and compare them with simulations, we assume constant values (ω-independent) for T_1, and for the phenomenological parameters associated with internal irreversibilities, friction work, and energy losses due to heat transfer. A reasonable set for these values is: $I_R = 1.4$ [34], $\varepsilon = 0.1$ [32], $\mu = 16\,\text{kg/s}$ [35]. Results for the efficiency and power output are plotted in Figs. 3.19 and 3.20, and for the different contributions to the total energy loss in Fig. 3.21. We compare in these figures, the theoretical FTT results with simulations results [2]. The geometrical parameters used for the calculations can be found in Appendix G and Table G.4.

The behavior of power and efficiency obtained from the theoretical model qualitatively agrees with those obtained from simulations (Fig. 3.19). Both curves, $\eta = \eta(\omega)$ and $P = P(\omega)$ show maximum values, the maximum of η is to the left to that of P. But differences are important. FTT-model predicts zero power output and efficiency for values of ω (above 500 rad/s) while simulated curves go to zero for angular speeds above 1400 rad/s. Moreover, the numerical values of both set of curves are quite different.

The existence of maxima in FTT curves for power and efficiency located at different speeds leads to loop-shaped power-efficiency curves, in accordance with simulation results (Fig. 3.20). These curves are obtained by parametric elimination of ω between $\eta = \eta(\omega)$ and $P = P(\omega)$. This kind of figures is very usual in

---

[5] Note that in this *irreversible* efficiency, η, the heat input, Q_{23}, is considered as coming from the reversible cycle. So the difference between the efficiency of the reversible, η_{rev}, and the corresponding irreversible cycle, η, comes only from the work output, calculated either in reversible, W_{rev} or irreversible basis, W. In both efficiencies, the heat input is considered the same.

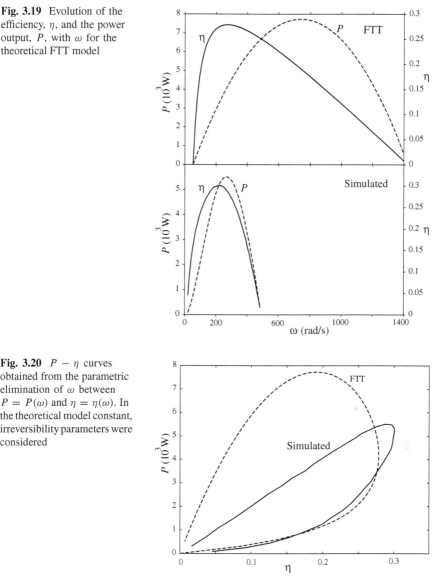

Fig. 3.19 Evolution of the efficiency, η, and the power output, P, with ω for the theoretical FTT model

Fig. 3.20 $P - \eta$ curves obtained from the parametric elimination of ω between $P = P(\omega)$ and $\eta = \eta(\omega)$. In the theoretical model constant, irreversibility parameters were considered

thermodynamic optimization,[6] because they allow to visualize the maximum power and maximum efficiency points, and the width of the region in between. This region is supposed to be the optimum one in the design of a wide variety of energy con-

---

[6] The parametric variable is not always the angular speed, it can be elected among the relevant variables of the system in which respect to its performance.

verters [12, 34, 36, 37]. But from Fig. 3.20, it is clear that the size and shape of the $P = P(\eta)$ curve obtained from FTT differs considerably from the simulated one.

In Fig. 3.21 are depicted the different contributions to work losses as calculated from simulations (lower panel) and from the FTT scheme [Eqs. (3.9), (3.14), and (3.16)]. In simulations, $|W_{rev}|$ is calculated from a computation without heat transfer to the cylinder walls, with no friction, and taking T_3 as the adiabatic flame temperature. $|W_I|$ is considered as the net work obtained from the simulations when friction forces and heat transfer to the cylinder wall are neglected. $|W_{fric}|$ is calculated directly from Eq. (3.14) and $|W_Q| = |W_I| - |W_{fric}| - |W|$. At sight of Fig. 3.21, the friction term qualitatively agrees in both frameworks, although discrepancies are found for heat transfer losses, and especially for the internal irreversibilities (solid lines in Fig. 3.21).

In consequence, FTT approach allows to reproduce some qualitative features in the behavior of power and efficiency with the engine speed, but a further improvement is required for a more precise theoretical description of the engine performance.

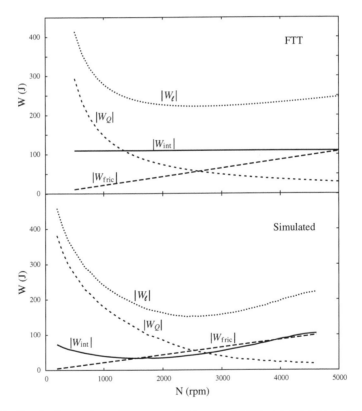

Fig. 3.21 Contributions to the total lost energy $|W_\ell| = |W_{int}| + |W_{fric}| + |W_Q|$ obtained within the theoretical FTT scheme or with quasi-dimensional simulations

We shall check now the possibility of improving the FTT model by considering as dependent on the speed some parameters that up to now were considered as constant.

It is expected that in a real engine the mass in the cylinder, m, and the inlet and combustion temperatures (T_1 and T_3 respectively), are functions of ω. Our procedure now will consist on using the simulations as a contrasted realistic method to calculate the functions $m = m(\omega)$, $T_1 = T_1(\omega)$, and $T_3 = T_3(\omega)$. Besides, some of the parameters that quantify in FTT the importance of irreversibility losses, I_R, and ε, that until this point were considered as speed independent and phenomenological, can be estimated as functions of ω from simulations. It will be shown that it is possible to get more realistic theoretical results using the simulation as a benchmark for the calculation of some ω-dependent parameters. Keeping the basic ideas and analytical character of FTT, results for the power output and efficiency, that are numerically closer to the real ones, will be obtained in the following subsection.

3.4.5 Refined FTT Approach with Speed Dependent Parameters

It is straightforward to calculate from simulations the mass inside the cylinder for any ω when the valves are closed. The plot in Fig. 3.22 shows a clear decrease of $m(\omega)$ from low revolutions up to almost half its value at high revolutions. This curve was obtained with simulations neither with heat nor friction losses.

With respect to the temperatures, T_1 (for each ω) is taken as the temperature when the piston is at BC before combustion and T_3 is taken as the temperature corresponding to an adiabatic isochoric combustion. The dependence of T_1 and T_3 on speed is displayed in Fig. 3.23. Note the monotonic increase of the inlet temperature T_1 from 360 to 430 K (about 20 %), while the adiabatic flame temperature T_3 shows a clear well-defined maximum value at intermediate revolutions. Now we deal with the calculation of the ω-dependence of the parameters I_R and ε.

In order to obtain a simple expression for the internal irreversibility parameter, I_R, Eq. (3.9) is expressed as,

$$|W_I| = T_3(\zeta_1 - I_R\zeta_2) \tag{3.17}$$

where:

$$\zeta_1 = \overline{C}_{v,23}\left(1 - \tau r^{\overline{\gamma}_{u,12}-1}\right)$$
$$\zeta_2 = \overline{C}_{v,41}\left(r^{\overline{\gamma}_{b,34}-1} - \tau\right) \tag{3.18}$$

Thus, if we denote, $|W_{\text{rev}}| = T_3(\zeta_1 - \zeta_2)$, it is possible to write I_R as:

$$I_R = \frac{\zeta_1}{\zeta_2} - \frac{|W_I|}{|W_{\text{rev}}|}\frac{\zeta_1 - \zeta_2}{\zeta_2} \tag{3.19}$$

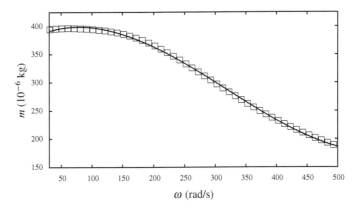

Fig. 3.22 Evolution of the mass contained in the cylinder with ω and the corresponding fitted polynomial

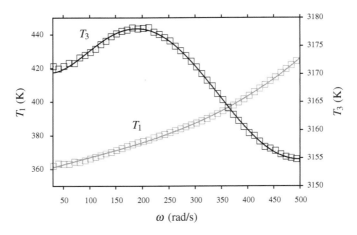

Fig. 3.23 Dependence of the inlet temperature, T_1, and the maximum cycle temperature, T_3 of the so-called *reversible* simulation

which will be ω-dependent because the temperatures, and therefore ζ_1 and ζ_2 are functions of the engine speed. We recall that $|W_I|$ is obtained from the simulations with no heat transfer energy losses and no friction losses. Moreover, $|W_{rev}|$ is calculated with the assumption that the combustion process is adiabatic and isochoric. So we are implicitly assuming that internal irreversibilities, $|W_{int}| = |W_{rev}| - |W_I|$, come from nonisochoric combustion, heat release processes during combustion, and also from pumping the gas mixture in and out the cylinder. Figure 3.24 shows a parabolic shape behavior for I_R with a minimum, $I_R = 1.15$, at 160 rad/s, and progressively increasing with ω. Then, a particular optimum regime is found where I_R is minimum while at higher velocities internal losses become quite large.

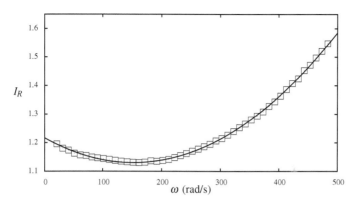

Fig. 3.24 Internal irreversibility factor, I_r as a function of ω obtained from the simulations and its corresponding fitting polynomial

The factor ε that relates heat transfer through the cylinder walls with work losses can be calculated by using Eqs. (3.6) and (3.16). Now ε will be a function of ω, because both T_3 and the temperature ratio τ depend on ω. Figure 3.25 shows a parabolic shape for ε with a maximum located at ω around 160 rad/s that is precisely the regime at which I_R has a minimum (Fig. 3.24). The interval of values of ε is quite wide, ranging from 0.04 at the extreme values of ω to almost 0.11 at its maximum value.

Numerical values of I_R and ε are similar to those found in the literature when considered as constant phenomenological parameters. Integral averages over all ω range of I_R and ε give us values of $\overline{I}_R = 1.23$ and $\overline{\varepsilon} = 0.08$ for the considered cylinder. These values compare favorably with the values 1.4 and 0.1 we took in Sect. 3.4.4 from previous works [23, 31, 34].

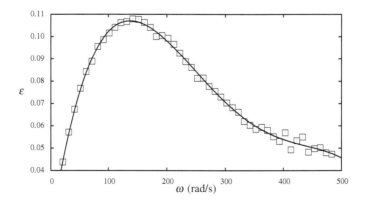

Fig. 3.25 Irreversibility parameter, ε, associated to heat transfer through cylinder walls and its corresponding fitting polynomial

Fig. 3.26 Comparison of
the evolution of the power
output and efficiency with the
rotational velocity as obtained
from simulations (*lines*) and
from the FTT theoretical
model (*squares*) including
ω-dependent polynomials

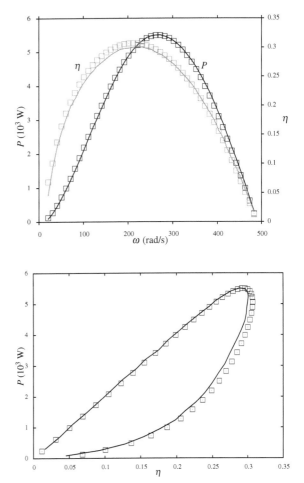

Fig. 3.27 Comparison of the
$P - \eta$ curves obtained from
the simulations (*solid line*) and
the FTT approach including
an ω dependence on some key
parameters

When polynomial fits to temperatures $T_1(\omega)$ and $T_3(\omega)$, mass inside the cylinder, $m(\omega)$, and the irreversibility parameters $I_R(\omega)$ and $\varepsilon(\omega)$ are incorporated, theoretical results for power and efficiency compare quite well, both qualitatively and quantitatively, with the simulated values for all the angular velocity (Fig. 3.26). Also, in the parametric $P - \eta$ curves, Fig. 3.27, good results are obtained from the theoretical model when the mentioned parameters are considered as ω-dependent as estimated from simulations. As a direct consequence, the different contributions to the lost work from FTT are now comparable to those in Fig. 3.21 from simulations: the minimum of $I_R(\omega)$ corresponds to a minimum in $|W_{int}|$ and the parabolic shape of $\varepsilon(\omega)$ produces an almost exponential decay of $|W_Q|$ as a function of ω.

In summary, a simple analytical model based on basic thermodynamic considerations and explicitly including chemical reactions and irreversibility losses associated with friction, heat transfer, and other internal losses is capable of reproducing detailed quasi-dimensional simulations for a single-cylinder Otto engine. By comparing both frameworks, interesting physical insight about the global irreversibilities related to power and efficiency losses is obtained. It will be shown in the next chapter that the combined applications of theoretical techniques and quasi-dimensional simulations can be very useful in engine design and optimization.

References

1. J. Heywood, *Internal Combustion Engine Fundamentals*, Chap. 3 (McGraw-Hill, New York, 1988), pp. 78–81
2. P.L. Curto-Risso, A. Medina, A. Calvo Hernández, J. Appl. Phys. **104**, 094911 (2008)
3. P.L. Curto-Risso, A. Medina, A. Calvo Hernández, J. Appl. Phys. **105**, 094904 (2009)
4. G. Beretta, M. Rashidi, J. Keck, Combust. Flame **52**, 217 (1983)
5. A. Fischer, K. Hoffmann, J. Non-Equilib, Thermodyn. **29**, 9 (2004)
6. J. Keck, in *Proceedings of Nineteenth Symposium (International) on Combustion* (The Combustion Institute, Pittsburgh, 1982), pp. 1451–66
7. H. Bayraktar, Renew. Energ. **30**, 1733 (2005)
8. M. Al-Hasan, Energ. Convers. Manage. **44**, 1547 (2003)
9. M. Ameri, B. Ghobadian, I. Baratian, Renew. Energy **33**, 1469 (2008)
10. W. Hsieh, R. Chen, T. Wu, T. Lin, Atmos. Environ. **36**, 403 (2002)
11. M. Celik, Appl. Therm. Eng. **28**, 396 (2008)
12. D. Descieux, M. Feidt, Appl. Therm. Eng. **27**, 1457 (2007)
13. F.L. Curzon, B. Ahlborn, Am. J. Phys. **43**, 22 (1975)
14. H.S. Leff, Am. J. Phys. **55**, 602 (1987)
15. A. Bejan, *Advanced Engineering Thermodynamics*, 3rd edn. (Wiley, Hoboken, 2006)
16. A. de Vos, *Thermodynamics of Solar Energy Conversion* (Wiley, Weinheim, 2008)
17. A. Durmayaz, O.S. Sogut, B. Sahin, H. Yavuz, Prog. Energ. Combust. **30**, 175 (2004)
18. S. Velasco, J.M.M. Roco, A. Medina, J.A. White, A. Calvo Hernández, J. Phys. D: Appl. Phys. **33**, 355 (2000)
19. B. Jiménez de Cisneros, L. Arias-Hernández, A. Calvo Hernández, Phys. Rev. E **73**, 057103 (2006)
20. B. Jiménez de Cisneros, A. Calvo Hernández, Phys. Rev. E **77**, 041127 (2008)
21. M. Esposito, R. Kawai, K. Lindenberg, C. Van der Broeck, Phys. Rev. Lett. **105**, 150603 (2010)
22. C. de Tomás, A. Calvo Hernández, J.M.M. Roco, Phys. Rev. E **85**, 010104(R) (2012)
23. F. Angulo-Brown, T.D. Navarrete-González, J.A. Rocha-Martínez, in *Recent Advances in Finite-Time Thermodynamics*, ed. by C. Wu, L. Chen, J. Chen (Nova Science Publishers, Commack, New York, 1999)
24. J.A. Rocha-Martínez, T.D. Navarrete-González, C.G. Pavía-Miller, A. Ramírez-Rojas, F. Angulo-Brown, Int. J. Amb. Energy **27**, 181 (2006)
25. S. Özkaynak, S. Götkun, H. Yavuz, J. Phys. D: Appl. Phys. **27**, 1139 (1994)
26. J. Chen, J. Phys. D: Appl. Phys. **27**, 1144 (1994)
27. X. Qin, L. Chen, F. Sun, C. Wu, Eur. J. Phys. **24**, 359 (2003)
28. Y. Ge, L. Chen, F. Sun, C. Wu, Appl. Energy **81**, 397 (2005)
29. A. Calvo Hernández, A. Medina, J.M.M. Roco, S. Velasco, Eur. J. Phys. **16**, 73 (1995)
30. A. Calvo Hernández, J.M.M. Roco, A. Medina, S. Velasco, Eur. J. Phys. **17**, 11 (1996)
31. J.M.M. Roco, A. Medina, A. Calvo Hernández, S. Velasco, Revista Española de Física **12**, 39 (1998)
32. M. Mozurkewich, R.S. Berry, J. Appl. Phys. **53**(1), 34 (1982)
33. Y. Ge, L. Chen, F. Sun, C. Wu, Int. J. Exergy **2**(3), 274 (2005)

34. F. Angulo-Brown, J.A. Rocha-Martínez, T.D. Navarrete-González, J. Phys. D: Appl. Phys. **29**, 80 (1996)
35. J. Heywood, *Internal Combustion Engine Fundamentals*, Chap. 13 (McGraw-Hill, New York, 1988), pp. 722–724
36. S. Sánchez-Orgaz, A. Medina, A. Calvo Hernández, Energ. Convers. Manage. **51**, 2134 (2010)
37. Y. Haseli, Energ. Convers. Manage. **68**, 133 (2013)

Chapter 4
Thermodynamic Engine Optimization

4.1 Design Parameter Optimization

The goal of this section is to analyze the sensitivity of a realistic Otto engine on two basic geometrical design parameters: the location of the ignition kernel with respect to the cylinder center and the stroke-to-bore ratio [1, 2], and to suggest possible optimum values. The parameters we analyze are just particular examples of variables that are easily managed in the simulations. In principle, other parameters such as those we mention at the end of this section can also be studied and optimized. Throughout this chapter, we shall consider the cylinder and valve geometry detailed in Appendix G.

4.1.1 Ignition Location

The *ignition location*, R_c, is the distance between the cylinder center and the position of the ignition kernel, where combustion begins to develop. When turbulence or fluctuations of the mixture around the spark plug are not considered, R_c represents the spark plug position. In Appendix C, we have presented an exact calculation of the flame front area, A_f, considered as spherical, for any value of R_c in terms of the cylinder geometry. Flame front area, and thus the evolution of combustion strongly depends on R_c. As R_c approaches the cylinder radius, the flame front reaches the cylinder wall, and thus its area is reduced. When R_c approaches zero, i.e., the spark plug is located close to the cylinder center, the flame front spends some more time to reach the walls, and so it develops with a larger area.

Figure 4.1 shows the results for the power output $P(\omega)$ and the efficiency $\eta_f(\omega)$ for three particular values of R_c: 0 (centered spark plug), 10 mm, and 20 mm.[1] From

---

[1] We recall here that the considered piston bore is 96 mm (see Table G.5 in Appendix G).

A. Medina et al., *Quasi-Dimensional Simulation of Spark Ignition Engines*, 87
DOI: 10.1007/978-1-4471-5289-7_4, © Springer-Verlag London 2014

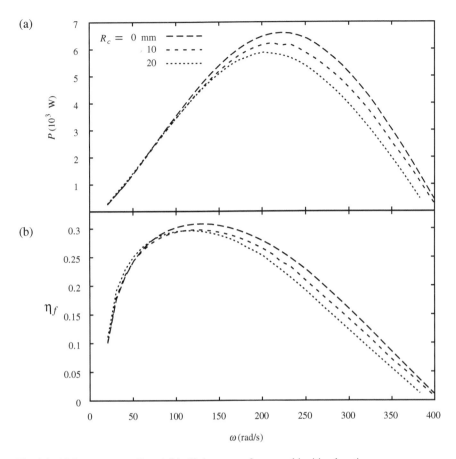

Fig. 4.1 (a) Power output, P, and (b) efficiency, η_f, for several ignition locations

the upper panel we see that at intermediate and high speeds, power output increases as R_c tends to zero with a right shifting of the corresponding maximum value.

The evolution of the engine efficiency with the ignition location is depicted in Fig. 4.1b. Similar to power output, for speeds over $\omega \simeq 100\,\text{rad/s}$, higher efficiencies are obtained as R_c decreases, while for very small speeds the evolution of efficiency with R_c is reversed: higher R_c gives higher efficiencies, but numerical differences are small. In Fig. 4.2, we have plotted power output versus efficiency by parametric elimination of the speed between the curves $P = P(\omega)$ and $\eta_f = \eta_f(\omega)$. The resulting loop-shaped curves are traversed counterclockwise as speed increases. An inspection of the curves obtained with different R_c values shows that a centered spark plug position $R_c = 0$ returns better values both for maximum power and maximum efficiency except for low speeds where the differences are small.

Quantitatively, for intermediate and high speeds, a centered spark plug position leads to power improvements around 10 % and efficiency improvements around 2 %.

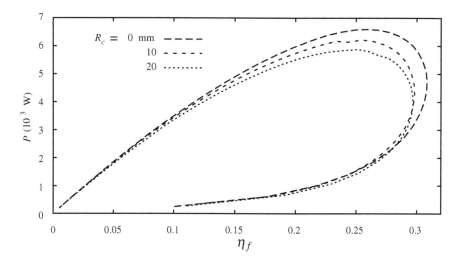

Fig. 4.2 Power-efficiency curves for different ignition locations, obtained by eliminating the speed, ω, between $P = P(\omega)$ and $\eta_f = \eta_f(\omega)$. Curves are built counterclockwise as speed increases

From a physical viewpoint, this could be associated with the size of the flame front area, that is larger for a centered spark and the combustion speed increases [3] (see Eq. (2.35), Appendix C, and Fig. C.3).

Two remarks are relevant at this point. First, the similar behavior of $P(\omega)$ and $\eta_f(\omega)$ with R_c observed in Fig. 4.1 can be explained by the relationship between power and efficiency on the basis of the chemical energy that enters the system through the fuel [$\eta_f = |W|/(m_{fuel}Q_{LHV})$], which is independent of the spark plug position R_c. Second, an increase in combustion speed leads to higher pressure inside the cylinder that could eventually exceed fuel auto-ignition limits. This could explain why in some real car engines the spark plug position is not centered in the cylinder [4]. Another reason to use noncentered spark plugs is to provide enough space in the cylinder head to place the valves. The simulations we are considering do not include auto-ignition effects, so this analysis should be used only for fuels with high enough octane number.

A complementary point of view to understand the influence of the spark plug position is to look at the behavior of the irreversibilities as they were defined in the previous chapter. In Fig. 4.3, we depict the work losses associated to heat transfer through cylinder walls, $|W_Q|$, friction losses, $|W_{fric}|$, internal irreversibilities, $|W_{int}|$, as well as the net work losses, $|W_\ell| = |W_{int}| + |W_Q| + |W_{fric}|$. From Fig. 4.3, the following features can be observed:

(i) Heat transfer, $|W_Q|$, and internal work losses, $|W_{int}|$, show at intermediate speeds a clear dependence on R_c, at intermediate speeds while they become practically independent of R_c at low and high speeds. Opposite, friction losses,

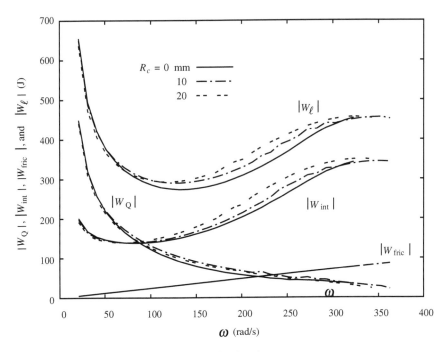

Fig. 4.3 Work losses analysis for different ignition locations

$|W_{\text{fric}}|$, are independent of R_c and according to the friction model elected for simulations [5], linearly dependent on ω.

(ii) At low speeds, net losses, $|W_\ell|$, are essentially due to heat transfer, but at intermediate and high speeds internal losses are dominant, so they rapidly decrease for low speeds as does $|W_Q|$, reach a minimum (approximately when $|W_{\text{int}}|$ becomes dominant), and then increase with ω.

(iii) Except for very low rotational speeds, the smallest total work losses $|W_\ell|$ are obtained with the centered ignition location. At intermediate speeds (between 100 and 300 rad/s), the net losses decrease as R_c decreases, in agreement with the observed increase of power and efficiency in this region.

4.1.2 Stroke-to-Bore Ratio

The *stroke-to-bore ratio* is the ratio of the stroke (twice the crankshaft radius, a) and the cylinder bore, B, $R_{\text{sb}} = 2a/B$. Small stroke-to-bore (*oversquare* engines, $R_{\text{sb}} < 1$) improves induction and exhaust (especially at high speeds) because it admits larger valves, but the combustion chamber has a poor surface-to-volume ratio resulting in more heat transfer losses [6]. On the other hand, piston friction losses are smaller because of the reduced distance traveled by the piston during each rotation. Crank stress is also lower due to lower peak piston speed relative to engine speed. This type of design is usually selected to obtain more peak torque at high speeds. *Undersquare* designs ($R_{sb} > 1$) favor fuel economy, because the better surface-to-volume ratio of the combustion chamber. For this reason, undersquare engines are increasingly being used in recent years. Variable stroke–length engines have been analyzed, either experimentally [7, 8] or by means of simulation analysis [9].

In this section, we analyze the influence of stroke-to-bore-ratio on the power output and efficiency, keeping the maximum volume of the combustion chamber, V_{cyl}, and the compression ratio, r, fixed. To do this, we use the following geometrical relations [10, 11]:

$$B = \left[\frac{4(r-1)V_{cyl}}{\pi r R_{sb}} \right]^{1/3} \tag{4.1}$$

$$a = \left[\frac{(r-1)V_{cyl}}{2\pi r} \right]^{1/3} R_{sb}^{2/3} \tag{4.2}$$

We consider the values of V_{cyl} and r given in Table G.5 (Appendix G) and change a and B to get an interval for R_{sb} between 0.2 and 1.5. Figure 4.4a represents the evolution of the power output with speed for four values of R_{sb}. At low speeds, the power output is almost independent of R_{sb}, but for ω over 100 rad/s, the influence of R_{sb} is important: the intermediate values of R_{sb} ($R_{sb} = 0.5$ and 1.0) return the highest power values, while low or high stroke-to-bore-ratios lead to smaller power output. The difference between the lowest one (corresponding to $R_{sb} = 0.2$) and the highest one (corresponding to $R_{sb} = 1.0$) is significant, around 18 %.

The effect of R_{sb} on efficiency, see Fig. 4.4b, is similar to that on power output: the intermediate values of R_{sb} give higher maximum efficiencies, although the relative differences (7 % between the outermost values) are smaller than for power output. From the results in Fig. 4.4, we emphasize that, for any R_{sb} value the speed giving the maximum power output could be considered as the operation limit of the engine, because for any speed over this value it is always possible to find another speed giving the same power but a better efficiency.

Simulations allow us to sweep all the interval for R_{sb} with a short step in order to analyze the possible nonlinear behavior of the maximum values of power output and efficiency in terms of R_{sb}, as seen in Fig. 4.4. Accordingly, in Fig. 4.5 the dependence of the maximum values of power output, P_{max}, and efficiency, $\eta_{f,max}$, with stroke-to-bore ratio, R_{sb}, are depicted. Actually, the evolution of both maxima is not linear: both curves have a maximum, which for power is around $R_{sb} = 0.8$ and for efficiency is around 0.6. Within the interval [0.6, 0.8], it is possible to simultaneously find high values of power output and efficiency. This would constitute the optimal working interval for this parameter when the objective is to obtain high values of power output and efficiency, independent of the speed.

Taking into account the fact that in a real engine, the power requirements change during its operation, a possible objective or optimization criterion is to obtain the

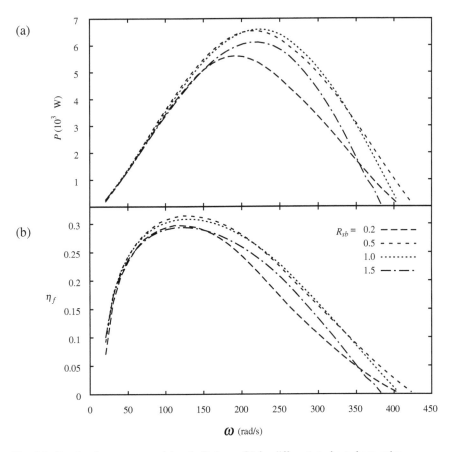

Fig. 4.4 Simulated power output (**a**) and efficiency (**b**) for different stroke-to-bore ratios

maximum efficiency for a required power output [1]. As shown in Fig. 4.6, when the required power is below approximately 2 kW, both small and high values of R_{sb} give very similar efficiencies, but for higher power requirements R_{sb} in the interval 0.5–1.0 leads to better efficiencies.

To understand the physical origin of the behavior of power and efficiency in terms of R_{sb}, we plot in Fig. 4.7 the evolution of work losses with engine speed for several values of R_{sb}. The differences among work losses associated with heat transfer through cylinder walls, $|W_Q|$, are only appreciable for speeds under 150 rad/s, and in this regime lower R_{sb}-values give more heat transfer losses as a consequence of the poor surface-to-volume ratio of the combustion chamber. As reported by Descieux et al. [12], for a compression ignition engine, the larger the stroke compared to the bore, the larger the mean velocity of the piston, and then friction losses, $|W_{fric}|$, are higher due to the longer distance traveled by the piston during each engine rotation.

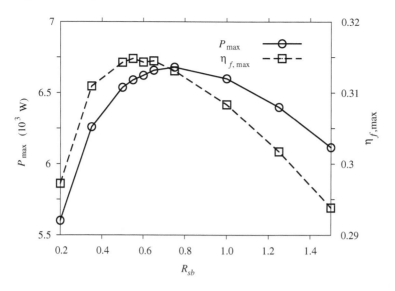

Fig. 4.5 Maximum values of the power output, P_{max}, and the efficiency, $\eta_{f,max}$, in terms of the stroke-to-bore ratio, R_{sb}

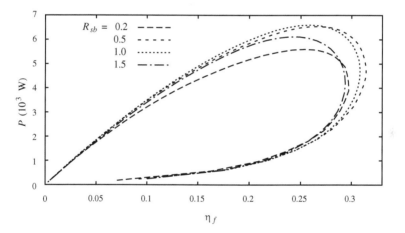

Fig. 4.6 Power-efficiency curves for several values of the stroke-to-bore ratio

Figure 4.7 also shows the expected results for friction losses: for a given R_{sb} they linearly increase with ω, and for a fixed ω linearly increase with R_{sb}.

Irreversibilities associated with fluid internal losses, $|W_{int}|$, are very sensitive to R_{sb}. In the speed interval between maximum efficiency and maximum power (approximately from 125 to 225 rad/s, Fig. 4.5), work losses are dominated by internal irreversibilities, $|W_{int}|$, which decreases with increasing R_{sb}, i.e., the opposite evolution that $|W_{fric}|$. Probably, this is associated with the larger time required by

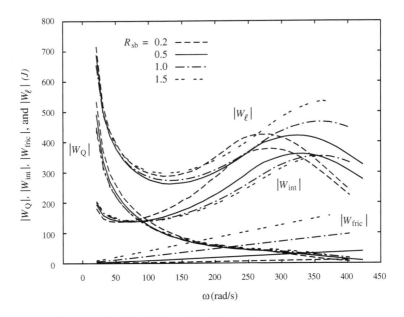

Fig. 4.7 Work losses as a function of the stroke-to-bore ratio

the flame to reach the cylinder walls for small R_{sb}, thus making combustion farther from an ideal isochoric process and promoting more internal irreversibilities. The combined influence of all these components, $|W_Q|$, $|W_{fric}|$, and $|W_{int}|$, leads to the minimization of the net work losses for R_{sb} values around 0.5–1.0 in the whole speed interval. This corroborates the optimal selection of this parameter as we discussed before.

Effects of other geometrical parameters such as the compression ratio, the chamber volume, or the valve geometry can also be investigated with quasi-dimensional simulations. The compression ratio is the ratio of the maximum and minimum volumes in the combustion chamber. Although it is not a pressure ratio, the relation is straightforward: as the compression ratio increases, the maximum pressure in the cycle also increases, leading to a larger power output. In order to avoid secondary effects, it is appropriate to make the corresponding analysis for a fixed chamber volume and for a fixed stroke-to-bore ratio. This can be done from Eqs. (4.1) and (4.2).

With respect to the chamber volume, there is a direct relationship between the cylinder volume and the mass (or the energy) that enters the chamber in each cycle, so its optimization is linked to fuel consumption. In this case, it is more convenient to use the *volumetric power output* (power output per unit volume) in order to optimize the engine design and to fix the other geometric parameters such as the stroke-to-bore ratio and the compression ratio.

4.2 Working Parameters Optimization

Now we analyze the sensitivity of a realistic spark ignition engine on three basic operation parameters: the spark advance, fuel–air ratio, and the internal wall temperature of the cylinder [1]. As before, we shall show their influence on the power output and efficiency. Possible optimization criteria and their physical interpretation in terms of the main types of work losses associated with thermodynamic irreversibilities will also be investigated.

4.2.1 Spark Advance

The location of the combustion event relative to the top center (TC) (which we take as 360°) is a basic parameter that it is often used to obtain the maximum achievable power or torque. Usually, combustion starts before the end of the compression stroke, its duration is between 30 and 90 crank angle degrees, and it finishes after the maximum pressure point in the cycle. If combustion is progressively advanced before TC, the work from the piston to cylinder gases during the compression stroke increases. On the other hand, if spark timing is retarded, pressure peak appears later in the expansion stroke and its maximum value decreases. As a consequence, useful work during the power stroke is reduced. Normally, *spark advance* angle is in the range between intake valve closure (around 220°) and TC. Conventionally, spark advance is optimized to give the maximum brake torque, called MBT timing [13], that takes place when the above-mentioned effects offset each other. Modern engines use computerized ignition systems, where a timing map with spark advance values for several combinations of engine speed and load. The computer sends a signal to the ignition coil at the indicated time in the timing map in order to fire the spark plug [14]. Our aim in this section is to propose an optimization method leading to a variable spark advance angle as a function of the engine speed.

From the simulations we show calculations of the power output, P, and the engine efficiency, η_f, for any rotational speed, ω, and for several values of spark advance, φ_0, i.e., we obtain $P = P(\omega, \varphi_0)$ and $\eta_f = \eta_f(\omega, \varphi_0)$. The behavior of these functions is plotted in Figs. 4.8 and 4.9 for different values of φ_0 between 300° and 350°. From these figures, it is possible to obtain the optimum value of ω that yields the maximum power and/or efficiency at any particular value of φ_0. By eliminating ω between $P(\omega, \varphi_0)$ and $\eta_f(\omega, \varphi_0)$ we can also generate the expected loop-shaped curves. These are displayed in Fig. 4.10.

We can go a step further by looking for an optimal spark advance with respect to both power and efficiency at each ω. Then for each ω, the values of φ_0 that gives the maximum reachable power and efficiency for that speed are found. If we denote these values by $\overline{\varphi}_0^P(\omega)$ and $\overline{\varphi}_0^{\eta_f}(\omega)$, respectively, a discrete number of pairs $(\omega, \overline{\varphi}_0^P(\omega))$ for power and a discrete number of pairs $(\omega, \overline{\varphi}_0^{\eta_f}(\omega))$ for efficiency are obtained. After a numerical interpolation we can obtain the functions $\overline{\varphi}_0^P = \overline{\varphi}_0^P(\omega)$ and $\overline{\varphi}_0^{\eta_f} = \overline{\varphi}_0^{\eta_f}(\omega)$.

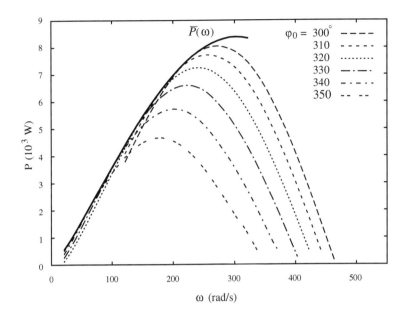

Fig. 4.8 Power output as a function of the rotational speed, ω, for different spark advance angles, φ_0. The envelope, $\overline{P} = \overline{P}(\omega)$, obtained by considering the optimized speed dependent ignition delay, $\overline{\varphi}_0(\omega)$, is also shown (*solid curve*). All the curves were obtained for a stoichiometric air–fuel mixture

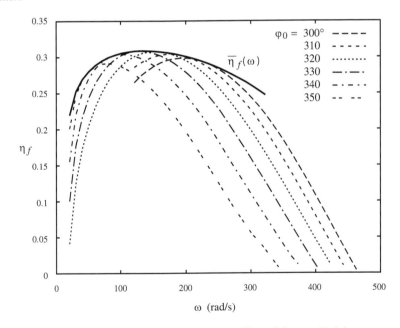

Fig. 4.9 As Fig. 4.8 but for the engine efficiency, η_f. The *solid curve*, $\overline{\eta}_f(\omega)$, represents the efficiency obtained with the optimum, speed dependent, spark advance, $\overline{\varphi}_0(\omega)$

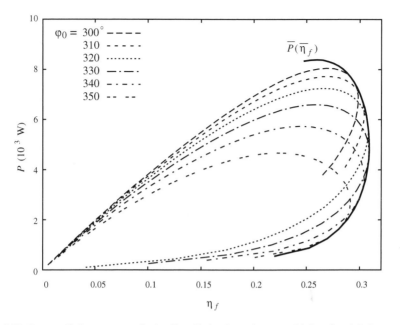

Fig. 4.10 Power-efficiency curves obtained by eliminating ω between $P(\omega)$ and $\eta_f(\omega)$ for several spark advance angles. The envelope, $\overline{P} = \overline{P}(\overline{\eta}_f)$, is obtained by taking the optimal speed dependent spark advance, $\overline{\varphi}_0(\omega)$

To a large extent, these functions are linear and very similar to each other. For the particular simulated engine we show in the figures, its numerical form is given by $\overline{\varphi}_0^P(\omega) \simeq \overline{\varphi}_0^{\eta_f}(\omega) \equiv \overline{\varphi}_0(\omega) = 363.85 - 0.27\,\omega$, with φ_0 in degrees and ω in rad/s.[2]

Finally, we can incorporate this ω-dependent spark advance in the power and efficiency formulations to obtain $\overline{P}(\omega) = P(\omega, \overline{\varphi}_0^P(\omega))$ and $\overline{\eta}_f(\omega) = \eta_f(\omega, \overline{\varphi}_0^{\eta_f}(\omega))$, which represent the envelopes depicted in Figs. 4.8 and 4.9. This is an important point, because it provides the possibility to obtain optimal spark advance values, as a function of engine speed, that simultaneously give maximum power and maximum efficiency. This is a reasonable result because, as shown in Eq. (3.4), efficiency is related to work (or power) through the mass of fuel inside the combustion chamber that is independent of spark advance; so optimization of engine efficiency or power should lead to the same result. The parametric plot, $\overline{P}(\overline{\eta}_f)$, is also shown in Fig. 4.10. For a given value of the power output (i.e., considering it as an exter-

---

[2] Note that in the limit $\omega \to 0$, the thermodynamic cycle developed by the engine corresponds to the Otto reversible cycle (performed in an infinite time), and in this case $\overline{\varphi}_0^P$ should be 360°, i.e., combustion begins when piston is at TC. In our case, this limit yields to a slightly different value due to the bi-parametric linear fitting procedure.

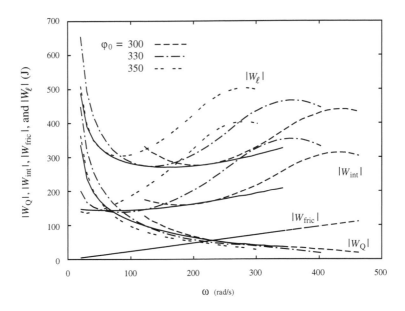

Fig. 4.11 Analysis of work losses for some spark advance angles. For each type of losses, those obtained with the optimized spark advance angle, $\overline{\varphi}_0(\omega)$, are shown in *solid lines*

nally fixed parameter), this envelope always gives the maximum reachable engine efficiency.[3]

The analysis of the optimization procedure in terms of work losses introduced in Sect. 3.4 leads to very interesting physical insights. Figure 4.11 shows the different terms composing the total work losses, $|W_\ell|$, for three values of φ_0 and also $\overline{\varphi}_0(\omega)$. From these plots, the following conclusions can be made:

(i) At low rotational speeds, total losses are dominated by heat transfer, but at intermediate and high speeds, the internal irreversibilities are the most important contribution to total losses. Friction losses are independent of φ_0, and increases monotonically with ω.

(ii) When the engine runs with optimal spark timing, $\overline{\varphi}_0(\omega)$ (solid lines), heat or internal losses are practically minimized: at low ω, those coming from heat losses, and at intermediate and high ω, those arising from internal irreversibilities. As a straightforward consequence, the addition of all losses, $|W_\ell|$, retains the minimum possible value at any rotational speed. Thus, the optimization procedure physically could be understood as a mechanism to minimize total losses by reducing heat transfer or internal irreversibilities losses as functions of the engine rotational speed in the different operation regimes of the engine.

---

[3] In this chapter, for optimization purposes, the fuel conversion efficiency is defined as the ratio between the power output per cycle and the chemical energy released. In addition, for fixed values of the chemical energy and the engine speed, maximum efficiency optimization corresponds to maximum brake torque optimization (MBT timing).

4.2.2 Fuel-Air Ratio

The *fuel–air (equivalence) ratio*, ϕ, is defined as the ratio of the actual fuel–air ratio to the stoichiometric value. Apart from affecting power output and engine efficiency, it has direct impact on pollutant emissions and autoignition. The fuel–air ratio should be close to unity for satisfactory spark ignition and flame propagation. Lean fuel–air mixtures (ϕ less than unity) will burn more slowly and reach lower maximum temperatures and also lower peak pressures, although they are beneficial in which respect to fuel consumption.

From the viewpoint of pollutant emissions, the engine-out emissions of spark ignition engines (mainly hydrocarbons, carbon monoxide, and nitrogen oxide) greatly exceed the levels required by most regulatory boards. There requirements can only be satisfied if appropriate exhaust gas after-treatment systems, as three-way catalytic converters, are used. Only for a very narrow fuel–air ratio *window*, whose mean value is slightly the stoichiometric value, can all the three pollutant species be almost completely converted to innocuous components (water and carbon dioxide). The mean fuel–air ratio can be kept within this narrowband only if electronic control systems together with the corresponding sensors and actuators are used. Fuel–air ratio control system is one of the most important control loops in spark ignition engines [14].

In this section, we focus on the relationship between fuel–air ratio and optimum engine performance. We shall explicitly see from simulations that it is possible to get an optimum fuel–air ratio as a function of rotational speed in order to maximize engine performance.

In Figs. 4.12 and 4.13, we have plotted power output, efficiency, and power versus efficiency curves obtained from quasi-dimensional simulations for different values of ϕ. The curves were obtained with the optimum variable spark angle function, $\overline{\varphi}_0(\omega)$, obtained in the previous section. While power increases at any value of ω with fuel–air ratio, the efficiency decreases at lower rotational speeds for ω approximately under 250 rad/s (note, in particular, that $\phi = 1.1$ gives the best results for power at any speed and the worst efficiency results at lower speeds). This is due to the presence of incomplete combustion products in the cylinder after combustion for rich mixtures.

At difference with spark advance, it is not straightforward to get an envelope for the curves $P(\omega, \phi)$ and $\eta_f(\omega, \phi)$ for different fuel–air ratios. However, it is possible to obtain a more subtle envelope as follows: for each power requirement we choose the value of ϕ giving the best engine efficiency and the corresponding speed. So we can perform a numerical interpolation to obtain the function $\overline{\phi} = \overline{\phi}(\omega)$, which for any fixed value of the power output gives the best achievable efficiency. As in the case of spark timing, this function is very simple, approximately linear, $\overline{\phi}(\omega) = 0.64 + 1.53 \times 10^{-3}\omega$, with ω in rad/s.

The result is the solid line in the parametric plot (Fig. 4.13). For instance, from the figure, it is obtained that for a power requirement of 7.5 kW, $\phi = 0.9$ gives $\eta_f = 0.26$, while for $\phi = 1.0$, $\eta_f = 0.29$. This represents an increase around 11 %. On the other hand, for a lower power requirement, a weaker mixture can lead to better efficiencies. The crossing points between $P = P(\eta_f)$ curves are also interesting.

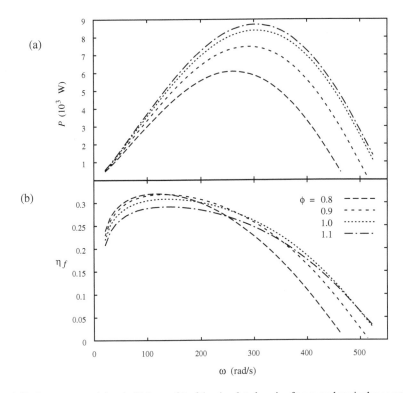

Fig. 4.12 Power output (**a**) and efficiency (**b**) of the simulated engine for several equivalence ratios, ϕ, ranging from lean to rich mixtures. Spark advance was considered as its ω-dependent optimal form, $\overline{\varphi}_0(\omega)$

For example, for $P \simeq 4\,\mathrm{kW}$, the simulations predict that for a mixture with $\phi = 0.8$ and other with $\phi = 0.9$ would lead to a similar efficiency, because there exists a crossing between the corresponding curves (see Fig. 4.13). The rotational speeds of the engine would be different depending on the fuel–air ratio, but the efficiency would be comparable in both configurations.

The influence of the fuel–air ratio on the losses associated with different irreversibility sources can be analyzed from Fig. 4.14. At low engine speed, the overall losses, $|W_\ell|$, are dominated by heat transfer to the cylinder walls, but when ω increases, these losses quickly diminish. Simultaneously, $|W_{\mathrm{fric}}|$ and $|W_{\mathrm{int}}|$ increase, so $|W_\ell|$ shows a minimum around $\omega \simeq 100\text{--}200\,\mathrm{rad/s}$ depending on the particular value of ϕ. It should be stressed that the overall losses increase with ϕ up to $\phi = 1.0$, where there is a change of behavior, and for $\phi = 1.1$, $|W_\ell|$ is lower. This effect can be understood in terms of $|W_{\mathrm{int}}| = |W_{\mathrm{rev}}| - |W_{\mathrm{I}}|$. $|W_{\mathrm{rev}}|$ is strongly dependent on the adiabatic flame temperature, T_3, that is very similar for $\phi = 1.0$ and $\phi = 1.1$, so $|W_{\mathrm{rev}}|$ is almost unaltered when passing through $\phi = 1.0$. But for $\phi = 1.1$, more fuel is inside the cylinder so more energy is released and $|W_{\mathrm{I}}|$ increases; thus, the

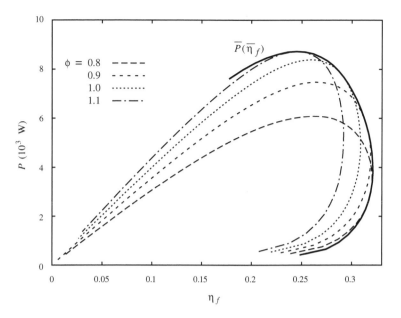

Fig. 4.13 Power-efficiency plots for several equivalence ratios with the same conditions as those in Fig. 4.12. The *solid line* is the curve obtained with the optimal ω-dependent fuel–air ratio, $\overline{\phi}(\omega)$, i.e., the fuel–air ratio giving for each required power output the maximum reachable engine efficiency

difference $|W_{int}|$ decreases. This reversal of the losses when passing through stoichiometric conditions is also responsible for the relative maximum found in $|W_\ell|$ around 250 rad/s when $\overline{\phi}(\omega)$ is considered (see the peak on the solid line for $|W_\ell|$ or $|W_{int}|$ in Fig. 4.14): total work losses increase up to reaching the curve corresponding to the largest losses ($\phi = 1.0$) and then decrease toward $\phi = 1.1$.

It should be mentioned that in contrast to spark advance, $\overline{\phi}(\omega)$, does not yield an envelope minimizing the net work losses at all rotational speeds, because, as noted before, this optimal fuel–air ratio does not lead simultaneously to envelopes of power and efficiency curves.

4.2.3 Wall Temperature

Heat transfer through the cylinder wall greatly affects engine performance and emissions. For a given rotational speed and a given mass of fuel inside the cylinder, an elevated heat transfer from the gas mixture to the cylinder wall decreases the average combustion temperature and pressure, thus the work per cycle transferred to the piston is reduced. In-cylinder, heat transfer occurs by convection and radiation, but for spark ignition engines radiation is usually considered to be negligible or subsumed into a convective heat transfer correlation [15]. In any case, heat transfer, as assumed

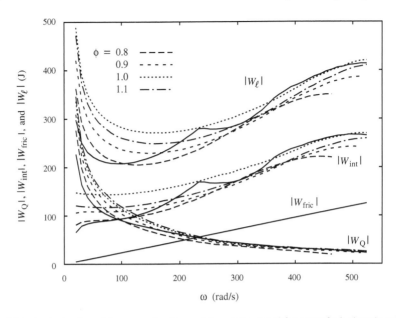

Fig. 4.14 Analysis of work losses as functions of the engine speed for some fuel–air ratio values (curves were obtained with the optimal, speed dependent, spark advance). Also shown, for each kind of losses, those arising from the optimized ω-dependent fuel–air ratio, $\bar{\phi}(\omega)$ (*solid lines*)

in Sect. 2.3.2, is proportional to the temperature difference $T - T_w$, where T is the instantaneous temperature of the bulk working fluid, and T_w, of the inner cylinder wall.

In Fig. 4.15, it is shown that the influence of the cylinder internal wall temperature, T_w, on the performance of the engine in the range of temperatures between 400 and 700 K, with the previously found optimal values of spark advance, $\bar{\varphi}_0(\omega)$, and fuel–air ratio, $\bar{\phi}(\omega)$. This figure shows a small but interesting dependence of power output and efficiency on T_w. At low angular velocities, both functions have higher values for low temperatures. But for intermediate values (between 240 and 310 rad/s), that is the region between maximum efficiency and maximum power points,[4] the situation is better analyzed by the parametric power versus efficiency curves illustrated in Fig. 4.16. A close inspection to this figure shows that in this region is always slightly better to keep a low internal wall temperature to get the maximum reachable efficiency for a particular fixed value of P.

The variation of work losses with T_w is depicted in Fig. 4.17 as a function of the angular speed. Friction losses and losses associated with internal irreversibilities do not depend on wall temperature, so only the influence of T_w is observed in heat transfer losses, $|W_Q|$. Because in the model assumed for the simulations, heat transfer linearly depends on the temperature gradient, $T - T_w$, it is expected that the heat

---

[4] Probably, the most adequate region for the stationary operation of any heat engine as argued by Chen [16–18].

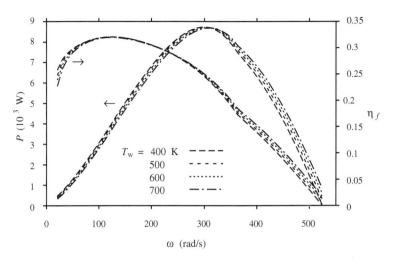

Fig. 4.15 Power and efficiency as functions of the engine speed for several cylinder internal wall temperatures, T_w. Spark advance and fuel–air ratio were considered at their ω-dependent optimal forms, $\overline{\varphi}_0(\omega)$ and $\overline{\phi}(\omega)$, respectively

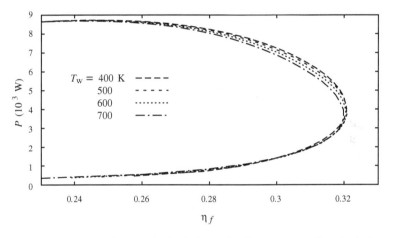

Fig. 4.16 $P = P(\eta_f)$ curves for several cylinder internal wall temperatures. Note that in the region between maximum efficiency and maximum power, the lowest temperature, $T_w = 400\,\mathrm{K}$, gives the best results: for a given power output it leads to the maximum achievable efficiency (because it is to the *right* of the other curves)

losses will increase with a decrease in T_w. This happens at high rotational speeds but the figure also shows that the opposite happens at low and intermediate values for ω (up to the point where there appears a crossing among curves, $\sim 310\,\mathrm{rad/s}$). This is associated with two linked effects: during intake, low wall temperatures reduce the heat transfer, thereby increasing the mixture density so more mass is charged per cycle; also, the mass flux increases and consequently the turbulent intensity

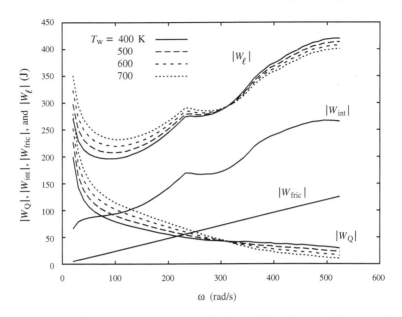

Fig. 4.17 Analysis of work losses components as functions of the engine speed for some values of T_w. All curves were obtained with the ω-dependent optimum spark advance and fuel–air ratios

increases thereby accelerating combustion process. So for speeds below 310 rad/s (approximately the point where engine power output is maximum), heat transfer work losses are minimal for the lowest T_w while beyond this speed the behavior of heat transfer losses is opposite.

4.2.4 Simultaneous Optimization of Working Parameters

Before we close this chapter, it is noteworthy to emphasize the capabilities of both numerical simulations and FTT tools for optimizing the performance of a spark igni-tion engine. To do this, we summarize in Fig. 4.18 and Table 4.1 the key results of each optimization step described above for the particular engine parameters considered in this chapter.

Starting from a simulated piston-cylinder system with constant spark advance angle, fuel–air ratio, for the reactants, and internal cylinder wall temperature, we have first studied the influence of different spark advance angles in terms of the rotational speed of the engine. It was concluded that for any power requirement, it is possible to find a spark ignition angle, as a function of speed, $\overline{\varphi}_0(\omega)$, giving the maximum reachable engine efficiency. In quantitative terms, the maximum power of the engine is increased around 27 % respect to the reference simulation. We also studied the influence of the fuel–air ratio, ϕ, of the unburned gases mixture over

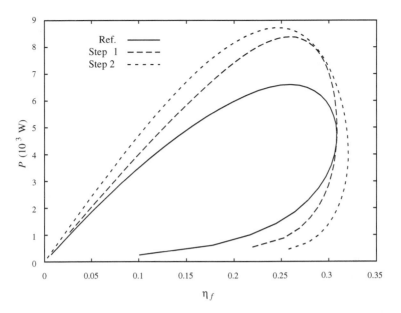

Fig. 4.18 Power-efficiency curves for the simulated engine obtained in the following way. Ref.: reference constant values for spark advance ($\varphi_0 = 330°$), fuel–air ratio ($\phi = 1.0$), and wall temperature, $T_w = 500$ K. Step 1: optimized ω-dependent spark advance ($\overline{\varphi}_0(\omega)$), constant fuel–air ratio ($\phi = 1.0$) and wall temperature, $T_w = 500$ K. Step 2: fully optimized values for the spark advance and fuel–air ratio ($\overline{\varphi}_0(\omega)$ and $\overline{\phi}(\omega)$, respectively) and wall temperature, $T_w = 400$ K. Notice that in this last step, the change in wall temperature from the reference value ($T_w = 400$ K) to the optimum one $T_w = 500$ K only introduces a small improvement in the region between maximum efficiency and maximum power conditions

Table 4.1 Values of maximum power, P_{max}, and maximum efficiency, $\eta_{f,max}$, in the three steps of the optimization procedure and values of the corresponding operating speeds, $\omega_{P_{max}}$ and $\omega_{\eta_{f,max}}$

	$\omega_{P_{max}}$ (rad/s)	P_{max} (kW) $[\Delta P_{max}(\%)]$	$\omega_{\eta_{f,max}}$ (rad/s)	$\eta_{f,max}$ $\left[\Delta\eta_{f,max}(\%)\right]$
Ref.	220.0	6.60	131.5	0.31
Step 1	302.4	8.39 [+27.1]	131.5	0.31 [– –]
Step 2	302.4	8.73 [+32.2]	121.5	0.32 [+3.2]

Ref.: reference constant values of spark advance, $\varphi_0 = 330°$, fuel–air ratio, $\phi = 1.0$, and $T_w = 500$ K. Step 1: optimal ω-dependent spark advance, $\overline{\varphi}_0(\omega)$, constant $\phi = 1.0$, and $T_w = 500$ K. Step 2: fully optimized values with ω-dependent spark advance, $\overline{\varphi}_0(\omega)$ and fuel–air ratio, $\overline{\phi}(\omega)$, and wall temperature, $T_w = 400$ K. The relative differences with respect to the reference case, $\Delta P_{max}(\%)$ and $\Delta\eta_{f,max}(\%)$ are shown in parenthesis

the already spark advance optimized engine. The main conclusion was that power output increases with ϕ for any rotational speed, but the effect on engine efficiency is more subtle because it depends on the engine speed. In any case, by analyzing the power-efficiency curves, it is also possible to derive a speed-dependent optimum fuel–air ratio, $\overline{\phi}(\omega)$, giving the maximum reachable efficiency for a certain power requirement. With the inclusion of this optimum fuel–air ratio in the simulations, we can obtain an additional improvement in the maximum power output around 4 %, and in efficiency around 3 %. Finally, the influence of the cylinder internal wall temperature, T_w, was analyzed. Although neither maximum power output nor efficiency are very sensitive to T_w, it is important to emphasize that globally the temperature giving the most favorable $P = P(\eta_f)$ is the lowest one, because it leads to better performance values in the region between maximum efficiency and maximum power values.

References

1. P.L. Curto-Risso, A. Medina, A. Calvo Hernández, J. Appl. Phys. **105**, 094904 (2009)
2. P.L. Curto-Risso, A. Medina, A. Calvo Hernández, Appl. Therm. Eng. **31**, 803 (2011)
3. M. Modarres Razavi, A. Hosseini, M. Dehnavi, *ASME Internal-Combustion-Engine-Division Fall Technical Conference 2009, Lucerne, Switzerland*, http://www.dsy.hu/thermo/razavi/Razavi_Spark2.pdf (2010)
4. C. Taylor, *The Internal Combustion Engine in Theory and Practice*, vol. I (The MIT Press, Cambridge, 1994)
5. P.L. Curto-Risso, A. Medina, A. Calvo Hernández, J. Appl. Phys. **104**, 094911 (2008)
6. R. Stone, *Introduction to Internal Combustion Engines* (Macmillan Press, London, 1999), Chap. 4, pp. 142–155
7. R. Siewert, SAE Trans. **87**, 3637 (1978)
8. F. Freudenstein, E. Maki, Variable-displacement piston engine, US Patent 4.270.495 (1981)
9. J. Yamin, M. Dado, Appl. Energ. **77**, 447 (2004)
10. J. Heywood, *Internal Combustion Engine Fundamentals* (McGraw-Hill, New York, 1988), Chap. 2, pp. 42–45
11. P.L. Curto-Risso, Numerical simulation and theoretical model for an irreversible Otto cycles. Ph.D. thesis, Universidad de Salamanca, Spain, http://campus.usal.es/gtfe/ (2009)
12. D. Descieux, M. Feidt, Appl. Therm. Eng. **27**, 1457 (2007)
13. J. Heywood, *Internal Combustion Engine Fundamentals* (McGraw-Hill, New York, 1988), Chap. 15, pp. 827–829
14. L. Guzzella, C.H. Onder, *Introduction to Modeling and Control of Internal Combustion Engine Systems* (Springer, Berlin, 2004)
15. R. Stone, *Introduction to Internal Combustion Engines* (Macmillan Press, London, 1999), Chap. 10, pp. 429–433
16. J. Chen, J. Phys. D: Appl. Phys. **27**, 1144 (1994)
17. A. Calvo Hernández, A. Medina, J.M.M. Roco, J.A. White, S. Velasco, Phys. Rev. E. **63**, 037102 (2001)
18. S. Sánchez-Orgaz, A. Medina, A. Calvo Hernández, Energ. Convers. Manage. **51**, 2134 (2010)

Chapter 5
Cycle-to-Cycle Variability

5.1 Models for Cyclic Variability

As it was detailed in the Introduction, Chap. 1, after decades of research on the cyclic fluctuations present in spark ignition engines, it is considered that the main causes of cyclic variability are: the variations in the motion of the gas mixture and the turbulences during combustion, the variations on the amount of the components of the gas mixture and its chemical composition, and the fluctuations of the peculiarities of the spark discharge, as the initial flame kernel position. Thus, any model susceptible to reproduce the experimental phenomenology of cyclic variability should incorporate at least some of those physical inputs. In Chap. 1 it is also possible to find a literature survey on different theoretical and simulation models in order to reproduce the main characteristics of cyclic variations.

In the type of quasi-dimensional simulations considered in this book, most of those physical ingredients are incorporated. The mass flow through the valves is calculated by means of the equations of a real mass flow through a restriction (Appendix B), assuming a discharge coefficient and the pressure gradient between the cylinder interior and the exhaust or intake valves. The quasi-dimensional combustion model by Keck and Beretta [1–3], Eq. (2.35) considers the presence of unburned eddies or vortexes of unburned gases within the flame front and is a turbulent combustion model. By means of a few parameters with a clear physical interpretation, this model allows to reproduce, as it was explained in Chap. 3, Sect. 3.3, the main characteristics of the evolution of the flame front in real spark ignition engines. Among those parameters it is the location of the initial flame kernel, so it is possible to analyze the influence of its variations in cycle-to-cycle variations. These simulation models also include explicitly an overlapping period in which intake and exhaust valves are simultaneously open, so the residual gases from the previous cycle mixes with the fresh air and fuel mixture. These residual gases are considered in the resolution of the chemical reactions associated with combustion, Appendix E. Moreover, the laminar flame speed of the actual cycle depends on the chemical composition of the residual gases from the previous one. This fact constitutes a kind of memory effect that links

A. Medina et al., *Quasi-Dimensional Simulation of Spark Ignition Engines*,
DOI: 10.1007/978-1-4471-5289-7_5, © Springer-Verlag London 2014

one cycle with the following one. Both effects lead to variations in the amount of fuel, air, and residual gases for each cycle as well as the chemical composition of the mixture. So as a global conclusion, it can be stated that most of the ingredients that are supposed to be in the origin of cyclic variability are readily incorporated in this kind of simulations.

5.2 Quasi-Dimensional Simulations and Cyclic Variability

As a summary of the chapter devoted to present the main theoretical aspects of quasi-dimensional models, Chap. 2 it is important to note that considering the engine as a dynamical system, the coupled ordinary differential equations for pressure and temperature are in turn coupled with two other ordinary differential equations for the evolution of the masses during combustion. So globally this leads to a system of differential equations where apart from several parameters (mainly arising from the geometry of the cylinder, the chemical reactions, and the models considered for other processes), there is a large number of variables: pressure inside the cylinder, p, temperatures of the unburned and burned gases, T_u, and T_b, and masses of the unburned and burned gases, m_u and m_b. All these variables evolve with time or with the crankshaft angle. So, up to this point, our dynamical model is a quite intricate deterministic system with those time-dependent variables. We shall see that this model itself is not capable to properly reproduce the cycle-to-cycle fluctuations observed in real engines in variables such as heat release, power output, efficiency, or others. But the basic model can be improved in order to solve this shortcoming. This will be the object of the present section.

Except when explicitly mentioned, we present numerical results obtained from quasi-dimensional simulations based upon the combustion model by Keck and Beretta described in Sect. 2.3.1. The numerical values for the cylinder geometry are those from Beretta et al. [2] at a fixed engine speed of 109 rad/s.

The computed results for heat release, Q_r, for each of the first 200 cycles are presented in Fig. 5.1a for a particular value of the fuel–air equivalence ratio, $\phi = 1.0$. Heat release is obtained as it was explained in Sect. 3.1. This time evolution was obtained directly from the solution of the deterministic set of differential equations. The curve clearly shows, after a transitory period, a regular evolution that does not match with previous experimental studies of cycle-to-cycle variability on heat release [4, 5] (see the inset in Fig. 5.1a). For other equivalence ratios, the curves obtained present different levels of variability but never display the typical experimental fluctuations. So the variability obtained by the direct solution of the nonlinear deterministic set of equations for temperatures and pressures, and the evolution of masses during combustion appears to be insufficient to reproduce the features of cyclic variability shown by experiments.

In order to analyze these time series we show in Table 5.1, for several fuel–air ratio values, some usual statistical parameters: the average value μ, the *standard deviation* σ, the *coefficient of variation*, $COV = \sigma/\mu$, the *skewness* $S = \sum_{i=1}^{N}(x_i - \mu)^3/$

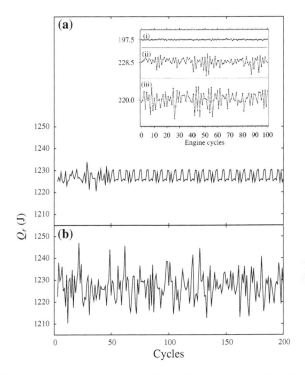

Fig. 5.1 (**a**) Heat release time series as obtained from the simulations without the consideration of any stochastic component for $\phi = 1.0$. The *inset* shows the experimental results obtained by Daw et al. [4] at three different fuel–air ratios: (i) $\phi = 0.91$, (ii) $\phi = 0.59$, and (iii) $\phi = 0.53$. The regular periodic simulated series obtained from deterministic simulations does not match with the experimental ones. (**b**) Heat release time series, Q_r, obtained with a log-normal distribution (see text) for the characteristic length of the unburned eddies during combustion, l_t for $\phi = 1.0$

$[(N-1)\sigma^3]$, and the *kurtosis* $K = \sum_{i=1}^{N}(x_i - \mu)^4/[(N-1)\sigma^4]$. S is a measure of the lack of symmetry in such a way that negative (positive) values imply the existence of

Table 5.1 Statistical parameters of the heat release temporal series, considering the deterministic model, for several fuel–air ratio values: mean value, μ, standard deviation, σ, coefficient of covariance, *COV*, skewness, S, and kurtosis, K

ϕ	0.5	0.6	0.7	0.8	0.9	1.0	1.1
μ (J)	500.84	746.19	886.82	1,019.73	1,138.79	1,227.20	1,253.32
σ (J)	2.593	0.320	0.785	1.463	1.969	2.249	2.255
COV ($\times 10^{-3}$)	5.177	0.428	0.885	1.435	1.729	1.832	1.799
S	−0.004	0.057	−0.106	−0.013	1.317	0.376	−0.062
K	2.485	2.638	1.332	1.326	3.610	1.361	3.458

Table 5.2 Statistical parameters of the heat release temporal series represented in Fig. 5.2 and obtained considering stochastic fluctuations in l_t, for several fuel–air ratio values

ϕ	0.5	0.6	0.7	0.8	0.9	1.0	1.1
μ (J)	492.50	726.23	885.43	1,020.24	1,139.51	1,227.16	1,253.71
σ (J)	122.966	57.872	17.937	7.333	6.326	6.663	7.093
COV ($\times 10^{-3}$)	250.0	79.7	20.3	7.2	5.6	5.4	5.7
S	−0.643	−3.028	−8.388	−3.194	0.147	0.423	0.335
K	2.495	13.455	104.475	49.491	3.579	3.710	3.454

left (right) asymmetric tails longer than the right (left) tail. K is a measure of whether the data are peaked ($K > 3$) or flat ($K < 3$) relative to a normal distribution.

With the objective to recover the experimental results on cyclic variability, we have checked the influence of incorporating a stochastic component in any of the physically relevant parameters of the combustion model. As detailed in Sect. 2.3.1, these parameters are the characteristic length of the unburned eddies entrained into the flame front, l_t and the characteristic speed u_t. Additionally, there is another basic parameter that influences the evolution of the flame front area and the heat transfer: the location of the initial flame kernel with respect to the cylinder center, R_c. We first analyze the influence of the characteristic length l_t and velocity u_t that give the evolution of burned and unburned masses during combustion, Eq. (2.35) keeping fixed the location of ignition at the spark plug position. Although both parameters could be considered as independent, we assume that they are linked by the empirical relations (2.42) and (2.45). There exist in the combustion literature experimental results for l_t. Particularly, we fitted the experimental results by Beretta et al. [2] considering l_t as a random variable with a log-normal probability distribution, $\mathrm{Log}N(\mu_{\log l_t}, \sigma_{\log l_t})$, around the nominal value $l_t^0 = 0.8\, L_{v,\max}(\rho_i/\rho_u)^{3/4}$ [Eq. (2.45)] with standard deviation $\sigma_{\log l_t} = 0.222$ and mean $\mu_{\log l_t} = \log(l_t^0) - (\sigma_{\log l_t}^2/2)$. Then, in each cycle u_t is obtained from Eq. (2.42) through the empirical densities ratio.[1]

For several values of ϕ ranging from very lean mixtures to over stoichiometric, Table 5.2 shows the characteristics of the time evolution of the heat release when the stochastic component on l_t has been introduced. Figure 5.1b displays a time series for the stochastic computations at least qualitatively much similar to the experimental ones in contrast to the deterministic for $\phi = 1.0$. It is important to note that it is difficult to perform a direct quantitative comparison with experiments, because of the large number of geometric and working parameters of the engine, usually not specified to the full extent in the experimental publications.

From a comparison between Tables 5.1 and 5.2, it is clear that average values are of course similar, but the standard deviation and coefficient of variation are much smaller in the deterministic case. The behavior of skewness is subtle, it only

---

[1] It is worth mentioning that the experiments by Beretta [2] measure the evolution of pressure during combustion with high-speed motion picture records of flame propagation. So we fit l_t with experiments on combustion and tried to recover cycle-to-cycle fluctuations by adding an stochastic behavior of a key parameter of combustion on the deterministic simulation.

Fig. 5.2 Representative sequences of heat release fluctuations for several values of ϕ. We observe evident changes in the mean and standard deviation of the signals as ϕ decreases. For intermediate values of ϕ, we identify the presence of multiple outliers corresponding to low values of heat release

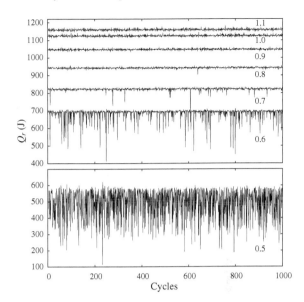

could be concluded that globally in the deterministic case values for S are closer to zero (that would be the value corresponding to a Gaussian distribution). The stochastic simulation kurtosis in the fuel–air ratio interval between $\phi = 0.6$ and 0.8 is much higher than in the deterministic case, that is a sign of more peaked distributions. Figure 5.2 shows the evolution of heat release with the number of cycles for fuel–air ratios[2] between 0.5 and 1.1. A careful inspection shows that at low and intermediate fuel ratios distributions are quite asymmetric with tails displaced to the left. In other words, there exist remarkable poor combustion or misfire events. This is quantified by the evolution with ϕ of the skewness compiled in Table 5.2: it takes high negative values in the interval $\phi = 0.5$–0.8. These results are in accordance with the experimental ones by Sen et al. [6] (see Fig. 1 therein).

We represent in Fig. 5.3, the evolution with the fuel–air equivalence ratio of the statistical parameters contained in Table 5.2 in order to perform a direct comparison with previous experimental results. Although the experiments by Daw et al. [4, 6] were performed for a real V8 gasoline engine with a different cylinder geometry that considered in our simulations, the dependence on ϕ of the standard deviation, coefficient of variation, skewness and kurtosis is very similar, although of course, vertical scales in simulations and experiments are different. It makes no sense to compare the mean values of heat release because of different shape and size of the real and the simulated engine. It is clear from the figure that at intermediate fuel–air ratios, around, $\phi = 0.7$, skewness also reproduces a pronounced minimum and kurtosis a sharp maximum. Nevertheless, the deterministic heat release computations

---

[2] An extrapolation of the laminar flame speed, S_L as a function of the fuel–air ratio, ϕ, was necessary for low values of ϕ.

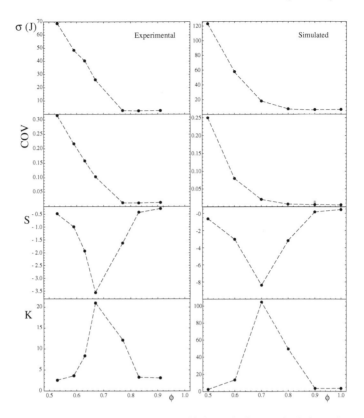

Fig. 5.3 Evolution of some statistical parameters with the equivalence ratio. *Left panel*, experiments [6] and *right panel*, stochastic simulation results

(statistical parameters in Table 5.2) do not reproduce neither the evolution with ϕ shown by experiments nor the magnitude of those statistical parameters.

Figure 5.4 represents the *first-return maps*, $Q_{r,i+1}$ versus $Q_{r,i}$, for the same values of ϕ obtained from the deterministic simulation (dark points) and also from the stochastic computations (color points). Return maps when l_t does not have a stochastic component, for any value of fuel–air ratio, seem noisy unstructured kernels. When l_t is considered as stochastic a rich variety of shapes for return maps is found, depending on the fuel–air ratio. From the stochastic return maps in this figure and the statistical results in Table 5.2 we stress the following points. First, at high values of ϕ (0.9–1.1) variations of heat release behave as small *noisy* spots characteristic of small amplitude distributions with asymmetric right tails and slightly more peaked near the mean than a Gaussian distribution. For $\phi = 1.0$ and 1.1, those noisy spots are partially overlapped. These results are very similar to the experimental ones by Green et al. for different engines (lower panels of Fig. 5.4) and those by Litak et al. (see Fig. 4 in [5]). Second, at intermediate fuel–air ratios ($\phi = 0.6$–0.8) extended *boomerang-shaped* patterns are clearly visible. Note that these series lead to

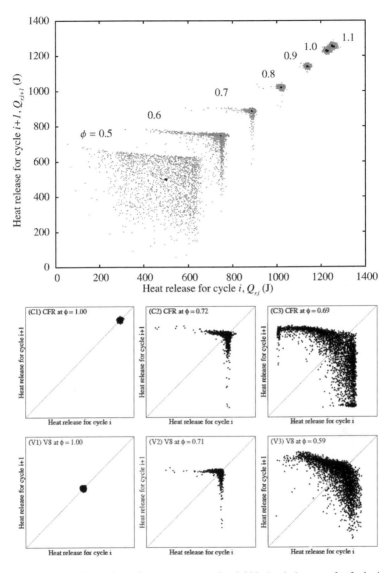

Fig. 5.4 *Upper panel* heat release first return maps after 2,800 simulation runs for fuel–air ratios between $\phi = 0.5$ and $\phi = 1$. l_t was considered as stochastic. For each fuel–air ratio, the central dark kernel corresponds to the deterministic simulation. *Lower panels* experimental heat release first return maps obtained by Green et al. [7]. The *upper row* of pictures (denoted CFR) corresponds to a single cylinder engine (Waukesha Model 48 CFR) and the *lower row* to a eight-cylinder production engine (Ford V8). Details can be found in [7]. Acronyms C1 and V1 refer to stoichiometric fueling, C2 and V2 to lean fueling, and C3 and V3 to very lean fueling

distributions that are very peaked near the mean and decline rather rapidly, and have quite asymmetric left tails. In this, interval simulation results also are in accordance with those of Daw et al. [4, 7]. Third, at the lowest equivalence ratio, $\phi = 0.5$ a change of structure is observed, probably because of cycle misfires for this very poor mixture. Here, the probability distribution is less peaked than a Gaussian, present a high standard deviation, and an asymmetric left tail. Fourth, a closer inspection of Fig. 5.4 reveals some kind of asymmetry around the diagonal, more pronounced as the fuel–air ratio becomes lower. This is in accordance with both experimental and model results by Daw et al. [4, 7, 8]. So an stochastic scheme with the consideration of pseudo-random fluctuations on l_t reproduces the characteristic return maps of this kind of systems from poor mixtures to over stoichiometric ones.

The results above refer to the joint influence of the parameters u_t and l_t on the CV-phenomena when both parameters are considered linked by the empirical relations (2.42) and (2.45). From now on our goal is twofold: on one side, to analyze the influence of u_t and l_t but considered as independent in the simulation model, and on the other side to analyze the influence of the third key parameter in the combustion description, the location of ignition accounted for the parameter R_c. To get this we shall show simulation runs for 2,800 consecutive cycles for the previously considered fuel–air equivalence ratios but taking these three parameters, one by one, as stochastic in nature.

First, we have considered only fluctuations on l_t, taking the same distribution as that in the beginning of this section. We show the corresponding return maps in Fig. 5.5a, that should be compared with Fig. 5.4. From this comparison we can conclude that the main characteristics of the return maps obtained when l_t and u_t were considered as linked (noisy spots near stoichiometric conditions, boomerang-shaped structures at intermediate fuel–air ratios, and complex patterns at lower air–fuel ratios) are already induced by the distribution of the characteristic length l_t when the other parameters are not stochastic. Differences between both figures only affect the dispersion of the boomerangs arms.

In regards to u_t, it is noteworthy that it is not easy to find in the literature experimental data which allow to deduce stochastic distributions for this characteristic turbulent velocity with certainty. We have used the data in [9] for the turbulence intensity and translated them to generate a log-normal distribution for the characteristic velocity, u_t, and checked slight changes in the distribution parameters within realistic intervals. So we assume a log-normal distribution $\mathrm{Log}N(\mu_{\log u_t}, \sigma_{\log u_t})$ around the nominal value $u_t^0 = 0.08\,\bar{u}_i(\rho_u/\rho_i)^{1/2}$ [Eq. (2.42)] with standard deviation $\sigma_{\log u_t} = 0.02\,u_t^0$ and mean $\mu_{\log u_t} = \log(u_t^0) - (\sigma_{\log u_t}^2/2)$. When u_t is the only stochastic parameter, the observed behaviors are not significantly altered (Fig. 5.5b): boomerang-like arrangements are found only at low ϕ, although some sensitivity in the arms of the boomerangs is appreciable. Moreover, the return maps do not reveal new features with respect to those generated with l_t.

Some authors [10] concluded from experiments and models that displacement of the flame kernel during the early stages of combustion has a major part in the origination of cycle-by-cycle variations in combustion. So it seems interesting to check this point from quasi-dimensional simulations. In our notation, R_c, represents

Fig. 5.5 Heat release return maps obtained when the three basic parameters of combustion, l_t, u_t, and R_c are considered as independent and one-by-one stochastic. (**a**) Fluctuations are only introduced in l_t (see text for details on the probability distribution); (**b**) only u_t fluctuates, and (**c**) only R_c is considered as stochastic

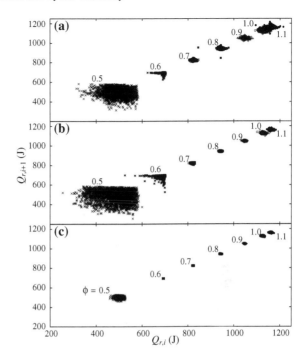

the position of the initial flame kernel relative to the cylinder center. The evolution of the flame front area (assuming an approximately spherical flame front) and the cylinder internal wall areas wetted by the enflamed volume (that play basic role in the calculation of heat transfers through cylinder walls) are directly influenced by R_c. According to the experimental results reported by Beretta [2], we assume a Gaussian distribution for R_c considered as an stochastic variable, with standard deviation $\sigma_{R_c} = 2.965 \times 10^{-3}$ m. The corresponding first-return maps are plotted in Fig. 5.5c. At sight of the figure, it is clear that the influence of a stochastic component in R_c is limited to cause noisy spots at any fuel–air ratio, clear structures as boomerangs are not found at any fuel–air equivalence ratio. So fluctuations on the displacement of the initial flame kernel does not seem sufficient to reproduce characteristic patterns of CV.

Finally, it has been checked what happens when the three parameters are simultaneously and independently introduced as stochastic in the simulations with the distributions mentioned above and no new features were discovered. In view of these results, we can say that the observed heat release behavior in terms of the fuel–air ratio is mainly due to stochastic and nonadditive variations of l_t and u_t and the nonlinearity induced by the dynamics of the combustion process, independently of ignition location fluctuations.

5.3 Nonlinear Analysis of Heat Release Fluctuations

5.3.1 Correlation Dimension Analysis

The simulation of very long time series of several engine variables by using quasi-dimensional combustion models is affordable from the viewpoint of the computation time required. After comparing the results obtained with experimental ones, we next explore the complexity of heat release fluctuations for different fuel–air ratio values. Particularly, we analyze in this section the dimensionality of the signals through the calculation of the *correlation dimension* [11–13]. This measure quantifies the self-similarity properties of a given sequence and is an important statistical tool to discriminate noise and determinism in the signal. We first reconstruct an auxiliary phase space by an *embedding* procedure. The *correlation sum*,

$$C(\varepsilon, m) = \frac{2}{N(N-1)} \sum_{i=1}^{N} \sum_{j=i+1}^{N} \Theta(\varepsilon - \| \mathbf{x}_i - \mathbf{x}_j \|) \tag{5.1}$$

where Θ is the *Heaviside function*, ε is a distance, and $\mathbf{x}_k$ are m-dimensional delay vectors, is computed for several values of m and ε [14]. If $C(\varepsilon, m) \propto \varepsilon^{-d}$, the correlation dimension is defined as $d(\varepsilon) = d \log C(\varepsilon, m)/d \log \varepsilon$. To get a good estimation of m it is required to repeat the calculations for several values of m, the number of embedding dimensions. This is the way to select which one corresponds to the dimensionality of the system. For low-dimensional deterministic signals, their phase space is typically characterized by a low *fractal dimension* value.

We apply the correlation dimension method to heat release sequences with 10^4 cycles and for different values of the fuel–air ratio in the interval $0.4 < \phi < 1.1$. According to the map-like characteristics observed in Fig. 5.5, we use a delay time equal to one. Figure 5.6 shows the correlation sum for selected values of the fuel–air ratio. We observe that for high and low values of ϕ ($\phi = 1.0$ and 0.4, Fig. 5.6a and c, respectively) the correlation integral shows a power law behavior against ε for several values of m. This behavior is best found by plotting the slope $d(\varepsilon)$ of $\log(\varepsilon, m)$ versus $\log \varepsilon$. The insets in each figure show these plots for values of m in the range 1–5. In both cases, there is a wide plateau of ε which corresponds to the power law behavior. It is also clear that the height of the plateau increases with the embedding dimension with a reduction of amplitude of the power law region and some fluctuations for low values of ε.

In contrast, we observe in Fig. 5.6b that for intermediate fuel–air ratios ($\phi = 0.65$), the scaling region only exists for values of ε in the range $10 < \varepsilon < 10^2$. This is confirmed by the presence of a plateau for several values of m (see the inset in Fig. 5.6b). Remarkably, the height of the plateau, $d(\varepsilon)$, almost does not change with the embedding dimension, indicating that the system can be characterized by a very low dimension. We also notice that for very small scales ($\varepsilon < 10$), a power law

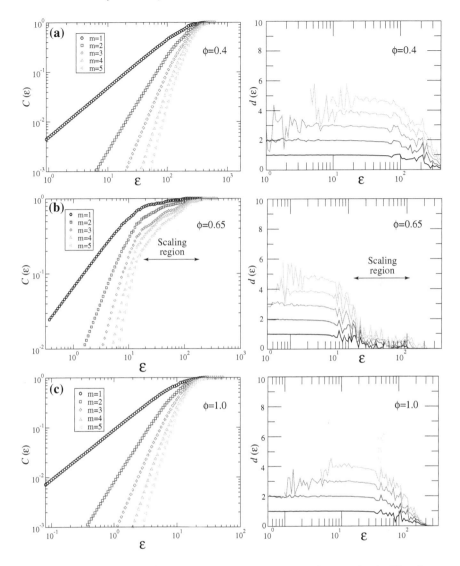

Fig. 5.6 Correlation integral, $C(\varepsilon)$, as a function of the distance, ε, for several embedding dimensions m and several values of ϕ. In each case, logarithmic plots of the correlation dimension, $d(\varepsilon)$, are also shown versus ε

behavior can be also identified, but the scaling exponent increases to reach the value of the embedding dimension, suggesting to identify this region as a noisy regime.

The findings discussed above are summarized in the plots of Fig. 5.7. For low and high fuel–air ratio values, the correlation dimension does not saturate for high embedding values, whereas for the intermediate fuel–air ratio shown ($\phi = 0.65$),

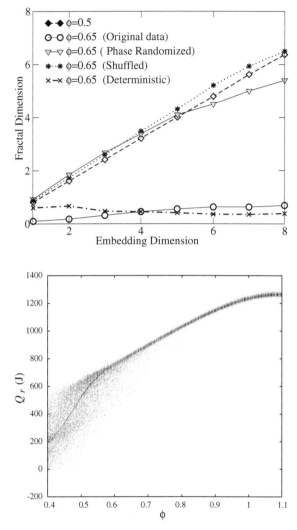

Fig. 5.7 Correlation dimension in terms of the embedding dimension for low and intermediate values of the fuel–air ratio

Fig. 5.8 Heat release, Q_r, time series as a function of ϕ. The *dark line* corresponds to the deterministic simulation, and the inhomogeneous dot distribution to the stochastic approach

correlation dimension saturates with the embedding dimension which is an indication of low-dimensional dynamics. It is also interesting that when fluctuations on l_t are not included in the simulations (deterministic case) the same situation is observed.

We present in Fig. 5.8 an extensive analysis of the evolution of the heat release with the fuel–air ratio between $\phi = 0.4$ and 1.1. The figure contains the results from the direct solution of the deterministic set of equations (dark line) and also from incorporating an stochastic component in l_t (dot distribution). For each value of ϕ, the last 100 simulated points of 250 cycles runs are shown. We remark that under moderately lean fuel conditions (ϕ around 0.65), our quasi-dimensional stochastic simulation does not show any period-2 bifurcation, as it happens in the theoretical

Fig. 5.9 Normalized heat release, $Q_r/\bar{Q}_r$ ($\bar{Q}_r$ is the average heat release for each fuel–air ratio) obtained from the deterministic simulations for several fuel–ratio, ϕ, intervals: (**a**) Whole fuel ratio range, from $\phi = 0.4$ to $\phi = 0.1$. (**b**) ϕ from 0.406 to 0.420. (**c**) ϕ from 0.570 to 0.585

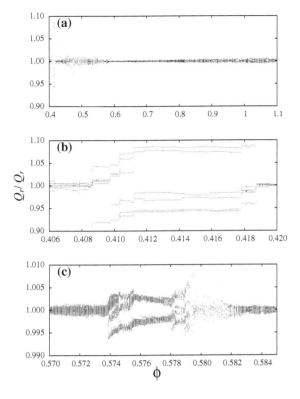

model by Daw et al. [4, 8]. Our work predicts a very low-dimensionality map in this region. Such conclusion makes sense since bifurcations have never been reported in real car engines.

On the other hand, the deterministic simulations do not lead to a fixed point as in the Daw's map, but to a variety of *multi-periodic* behavior for different fuel–air ratios. In particular, Fig. 5.9 shows the evolution of the normalized deterministic heat release, $Q_r/\bar{Q}_r$, ($\bar{Q}_r$ is the average heat release for each value of ϕ) with fuel–air ratio between 0.4 and 1.1. In Fig. 5.9a, we observe that for low and high fuel–air ratio values the number of fixed points is clearly smaller than those present for intermediate values of ϕ, indicating a high multi-periodicity. It is difficult to state (even with very long simulations) if the patterns observed around $\phi = 0.42$ and $\phi = 0.57$ could be considered as period-2 bifurcations, because of their scale dependence. In the zooms (Fig. 5.9b, c), there seem to be two regions with a large concentration of fixed points separated by a very small vertical distance (note the different scales of vertical axes of the plots in Fig. 5.9), but when we see Fig. 5.9b, c from apart those two regions resemble something similar to two large dot clouds. It is worth mentioning that both patterns appear in regions just before the density of fixed points begin to increase. In any case, these patterns disappear when stochastic components in any of the main combustion parameter are introduced in the simulations which is in accordance with the absence of heat release bifurcations in real engines. A deeper evaluation of the multi-periodicity observed in bifurcation-like plots for this system probably will deserve future studies.

To remark the findings about dimensionality of heat release sequences, we recall that the correlation dimension of the dynamical system, obtained through a reconstruction of the phase space from heat release time series is small only for intermediate fuel–air ratios, where the deterministic system and the stochastic one have almost the same dimensionality. When a systematic study of the evolution of the heat release calculated from the stochastic simulations with ϕ is performed, bifurcations were not found. This is in agreement with real engines.

5.3.2 Monofractal Analysis

One key question when analyzing noisy-fluctuating data is to evaluate the presence of correlations. For instance, the *power spectrum* is the typical method to characterize autocorrelations in time series. For stationary stochastic process with *autocorrelation function* which follows a power law $C(s) \sim s^{-\gamma}$ where s is the lag and γ is the correlation exponent, $0 < \gamma < 1$, the presence of long-term correlations is related to the fact that the mean correlation time diverges for infinite time series. According to the *Wiener-Khinchin theorem*, the power spectrum is the Fourier transform of the autocorrelation function $C(s)$ and for the case described above, we have the scaling relation $S(f) \sim f^{-\beta}$, where β is called the *spectral exponent* and it is related to the *correlation exponent* by $\gamma = 1 - \beta$.

Alternative methods have been proposed to the assessment of correlations and fractal properties for stationary and nonstationary time series [15–17]. In [17], Higuchi proposed a method to calculate the *fractal dimension* (FDM) of self-affine curves in terms of the slope of the straight line that fits the length of the curve versus the time interval (the lag) in a double log plot. The method consists in considering a finite set of data taken at an interval $\nu_1, \nu_2, \ldots, \nu_N$. From this series we construct new time series, ν_m^k, defined as

$$\nu(m), \nu(m+k), \nu(m+2k), \ldots, \nu\left(m + \left[\tfrac{N-k}{k}\right] \cdot k\right); \text{ with } m = 1, 2, 3, \ldots, k. \tag{5.2}$$

where [ ] denotes Gauss' notation, that is, the bigger integer and m and k are integers that indicate the initial time and the interval time, respectively. The length of the curve v_m^k, is defined as:

$$L_m(k) = \frac{1}{k} \left[\left(\sum_{i=1}^{\left[\frac{N-m}{k}\right]} |\nu(m+ik) - \nu(m+(i-1)k)| \right) \frac{N-1}{\left[\frac{N-m}{k}\right] k} \right] \tag{5.3}$$

and the term $(N-1)/[(N-m)/k]$ represents a normalization factor. Then, the length of the curve for the time interval k is given by $\langle L(k)\rangle$: the average value over k sets $L_m(k)$. Finally, if $\langle L(k)\rangle \propto k^{-D}$, then the curve is fractal with dimension D [17].

The fractal dimension is related to the spectral exponent β by means of $\beta = 5 - 2D$ [17]. Note that this relationship is valid for $1 < D < 2$ and $1 < \beta < 3$. For *uncorrelated random walk*, which results from the integration of *white noise* fluctuations, we observe $D = 1.5$. For β within the interval $-1 < \beta < 1$, that is, for processes which can be described as the first derivative of fluctuations with spectral exponent within the interval $1 < \beta < 3$, the relationship between β and D changes to $\beta = 3 - 2D$. A process with *positive long range correlations* leads to $D < 1.5$, whereas for *anti-correlated processes* $D > 1.5$.

First, we apply the FDM to cyclic heat release fluctuations obtained from simulations according to the model described above. A direct application of the FDM to heat release sequences showed in Fig. 5.2 reveals that the scaling behavior is represented by exponents close to 2, which correspond to white noise fluctuations. However, FDM becomes highly inaccurate for anti-correlated signals, especially for sequences with spectral exponent within the interval $-1 < \beta < 1$ [17, 18]. For a more reliable application of the FDM to the heat release data, the sequence, $Q(i)$, is first integrated to obtain the profile $Q_I(i) = \sum_{j=1}^{i}(Q_j - \bar{Q})$, with $\bar{Q}$ the mean value. In this way, the integrated signals lead to fractal dimensions within the vicinity of $D = 1.5$, where the FDM has been proved to estimate a stable scaling exponent [17–19]. For instance, a white noise sequence leads to a fractal dimension value of $D = 1.5$ and corresponds to the case where events are not correlated. For $D > 1.5$, the fluctuations resemble anti-persistent behavior whereas for $D < 1.5$, the variations are described as persistent with long range correlations. Figure 5.10 shows representative cases of $\langle L(k)\rangle$ versus k for integrated heat release sequences and several values of the fuel–air ratio. We observe that for high and low values of ϕ, the statistics follows a power law behavior with an exponent slightly larger than $D = 1.5$ (see next paragraph for a quantitative evaluation), indicating that heat release values

Fig. 5.10 Plot of log $\langle L(k)\rangle$ versus log k of heat release sequences for several values of the fuel–air ratio ϕ

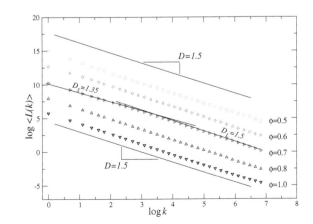

are close to uncorrelated variations with a weak *anti-persistence*, that is, heat release values tend to alternate. In contrast for intermediate values of the fuel–air ratio, we identify two regimes separated by a crossover point.

In order to estimate the D-values and the crossover scale, we consider the following procedure: given the statistics of $\langle L(k)\rangle$, a sliding pointer along k is considered to perform linear regression fits to the values on the left and to the elements on the right. At each position of the pointer, we calculate the errors in the fits (e_l and e_r) and monitor the total error defined by $e_t = e_l + e_r$. We define two stable exponents (D_S and D_L) when e_t reaches its minimum value and the position of the crossover point is within the interval $10 \leq k \leq 500$. The results of this analysis are depicted in Fig. 5.11, we observe that for intermediate values of ϕ, over short scales the fractal dimension is smaller than $D_S = 1.5$, whereas for large scales the exponent D_L is close to the white noise value. It is important to remark that for both low and high fuel–air ratio values, the scaling exponent ($D \gtrsim 1.5$) is identified with a weak anti-persistent behavior with anti-correlated fluctuations whereas at intermediate values, the scaling exponent for short scales is quite different (smaller) than the uncorrelated value $D = 1.5$, revealing that at small scales the fluctuations are correlated with a weak persistent behavior, in the sense that an increment of the heat release value is more likely to be followed by an increment and the same occurs for decrements.

Figure 5.11 also shows how around $\phi = 0.65$, D_S reaches a kind of minimum value. For this case, the crossover point which separates short and large regimes is located around $k^* \simeq 20$ cycles. According to Heywood [20], engine performance measurements have showed that the engine stability limit evidenced by a minimum fuel consumption and onset of rapid increase in hydrocarbon emissions occurred at $\phi = 0.65$, just before the partial burn limit where some slow-burning cycles occur but combustion is still complete in all cycles. That is, around $\phi \simeq 0.65$ is the border between complete combustion cycles and partial burn until values of $\phi \simeq 0.5$ where

Fig. 5.11 Statistics of D_S and D_L for several values of the fuel–air ratio ϕ. For intermediate values of ϕ, two scaling regimes are identified separated by a crossover point. For short scales we observed $D_S < 1.5$, whereas for large scales $D_L \simeq 1.5$

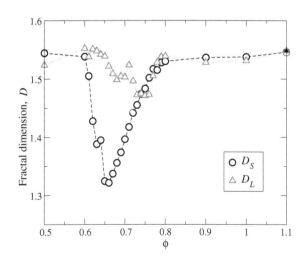

many misfires occur. Besides, around $\phi \simeq 0.65$, the quasi-dimensional numerical model here studied produces combustion heat time series corresponding to a low-dimensionality dynamics [21], which represents deterministic behavior of CV, which according to Scholl and Russ [22] is consequence of incomplete combustion. On the other hand, in Fig. 5.11 we observe that for low and high values of ϕ, both D_S and D_L are equal or bigger than 1.5 corresponding to dynamical regimes close to white noise of fluctuations with anti-persistent behavior.

From the perspective of correlations based on a monofractal approach, we have analyzed heat release sequences for different fuel–air ratio values. Our results show that there are important changes in the dynamical behavior as ϕ changes, particularly, transitions from a weak anti-correlated to a positive correlated behavior for short scales. In order to further evaluate the presence and origin of correlations in the signals, we repeated our calculations with *surrogated sequences* obtained from random permutations of the heat release values and from a *phase randomization* of the sequence. For the first case, the original sequence was randomly shuffled to destroy temporal correlations but preserving the distribution of events. For the second case, the surrogate set was constructed by a phase randomization of the original sequence. In this way, nonlinearities are eliminated but preserving the original power spectrum and changing the distribution. The phase randomization was performed after applying the Fast Fourier Transform (FFT) algorithm to the original time series to obtain the amplitudes. Then the surrogate set was obtained by applying the inverse FFT procedure [23, 24]. Figure 5.12 shows the results of fractal dimension for shuffled and phase randomized data. For shuffled data, the values of D_S and D_L are close to the values of original data (Fig. 5.11), while phase randomized data are characterized by a single fractal dimension with values close to $D = 1.5$, indicating an uncorrelated behavior even for intermediate values of the fuel–air ratio.

Fig. 5.12 Statistics of D_S and D_L for shuffled and phase randomized heat release sequences. Phase randomized is described by one scaling exponent which is close to the white noise value, whereas shuffled data lead to values which are close to those from original simulated data (Fig. 5.11)

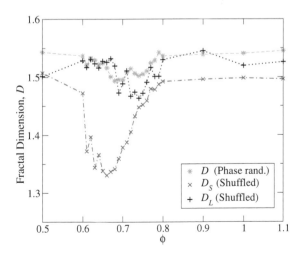

5.3.3 *Multifractal Analysis*

In many cases, we deal with signals which exhibit a complex behavior which is not fully characterized by a single scaling exponent. For this type of records, a larger number of exponents are necessary to characterize their scaling properties. Multifractal signals can be analyzed at many scales with different local scaling exponents (*Hurst exponents*, h). The local Hurst exponent quantifies the singular behavior for a given scale. For monofractal signals, the local exponent is the same for the entire signal ($h = H$), with H the *global Hurst exponent*. It is known that for positively correlated signals the value of H ranges from $1/2$ to 1 ($1/2 < H < 1$), whereas for anti-correlated processes $0 < H < 1/2$. In fact, for monofractal signals there is a relationship between H and the fractal dimension of the time series D, given by $D = 2 - H$ [25, 26]. Time series with multifractal properties are characterized by a set of local exponents h which conform the function $f(h)$, where $f(h)$ represents the fractal dimension of the subset of the signal characterized by the local exponent h.

To determine the values of the local exponents h, we apply the *wavelet multifractal method* (WMM) which is very suitable for the analysis of multifractal signals [27–29]. The wavelet transform is very appropriate because it can remove local trends, reducing the effects of nonstationarities. Specifically, we use the *wavelet transform modulus maxima method* to calculate h. The third derivative of the Gaussian function was used as wavelet function in all our calculations [27, 28]. First, we consider a partition function $Z_q(a)$ defined as the sum of the q-th powers of the local maxima of the modulus of the wavelet transform coefficient at scale a. For small scales, a power law behavior is expected between $Z_q(a)$ and a, that is, $Z_q(a) \sim a^{\tau(q)}$, with $\tau(q)$ a exponent which characterizes the scaling behavior at different scales. For positive q, the statistics reflects the scaling of large fluctuations and strong singularities, while for negative q, small fluctuations and weak singularities. The exponent $\tau(q)$ is related to the Hurst exponent through the relationship $\tau(q) = qh - 1$, where $h = d\tau(q)/dq$. Note that for monofractal signals $d\tau(q)/dq = H$, with H a constant for the entire signal. The *singularity spectrum*, $f(h)$, is obtained by means of the Legendre transform given by, $f(h) = q \, d\tau(q)/dq - \tau(q)$. We characterize the multifractality properties of signals by $\Delta h = h_{\max} - h_{\min}$, which represents the width of the singularity spectrum and h^* which is the value that maximizes $f(h)$. When $h^* = 0.5$, we refer to an uncorrelated random walk; whereas for $h^* > 0.5$, a process dominated by fluctuations with positive long range correlations and for $h^* < 0.5$, a process with anti-persistent correlations.

To get further insight in the analysis of cyclic fluctuations, we apply the multifractal method to the integrated variations of heat release. Figure 5.13a shows the behavior of $\tau(q)$ versus q for several values of ϕ from lean to stoichiometric conditions. We observe that for intermediate values of the fuel–air ratio, a nonlinear behavior between $\tau(q)$ and q is identified, revealing multifractal properties of the signal. Figure 5.13b shows that, for intermediate values of ϕ, the singularity spectrum is broad, confirming the multifractality of the signal. In contrast, for high and low fuel–air ratio values, an almost linear relationship between $\tau(q)$ and q is identified

Fig. 5.13 Multifractal analysis of heat release variations. (a) Multifractal spectrum $\tau(q)$ for several values of ϕ. For intermediate values of ϕ, a nonlinear relationship is identified indicating multifractality. (b) Singularity spectrum $f(h)$ versus h for the cases showed in (a)

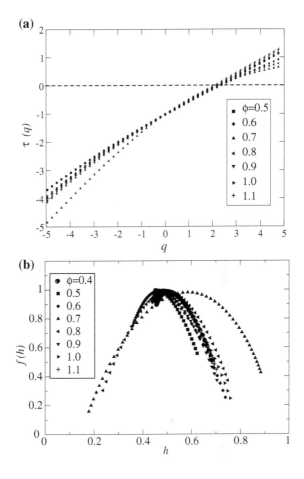

(Fig. 5.13a), which indicates a monofractal property as it can be seen in Fig. 5.13b, where $f(h)$ is described over a narrow range of exponents h. The statistics of the singularity spectrum further reveals that for intermediate values of ϕ, the local Hurst exponents cover a broad range related to multifractality (see Fig. 5.13a). Interestingly, we observe that the highest degree of multifractality is reached for values around $\phi \simeq 0.68$.

The statistics of the local exponent h^* which maximizes the singularity spectrum is summarized in Fig. 5.14b. From this figure, it is clear that there is a transition for h^* from $h^* < 0.5$ to $h^* > 0.5$ as the fuel–air ratio varies from $\phi = 0.5$ to 0.65; and vice versa from $\phi = 0.78$ to 0.8, revealing that for intermediate values of the fuel–air ratio, the fluctuations are dominated by a positive correlated behavior, whereas for high and low fuel–air ratio the dynamics can be described as processes with antipersistent variations. We also observe that the maximum value of h^* occurs when $\phi \simeq 0.67$ which roughly corresponds to the value where the fluctuations reach the

Fig. 5.14 (a) Width of
the singularity spectrum
$\Delta h = h_{\max} - h_{\min}$ versus ϕ.
A broad spectrum represents a
high degree of multifractality.
(b) Statistics of h^* versus ϕ.
For $h^* > 0.5$, the process is
dominated by positive correlations whereas for $h^* < 0.5$,
the process is described as
anti-correlated. We observe
that for intermediate values
of ϕ, the dominant dynamics
can be described as positively
correlated

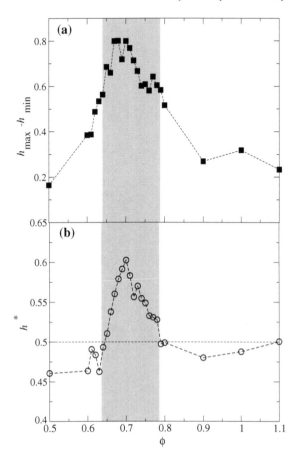

highest degree of multifractality (Fig. 5.13b). All these results are consistent with the
results of the FDM analysis of previous section.

Multifractal analysis reveals that shuffled data are close to original series while
phase randomized data exhibit a narrow spectrum close to a monofractal behavior
(Fig. 5.15a). Figure 5.15b shows the results of h^* versus ϕ, revealing that phase randomized data exhibit uncorrelated fluctuations with values around $h^* \simeq 0.5$, while
for shuffled data, the dominant dynamics for intermediate values of the fuel–air ratio
is quite similar to that observed in original data (Fig. 5.14b). The results from shuffled and phase randomized sequences for intermediate fuel–air ratios suggest that
phase correlations are important, because when they were eliminated the fluctuations exhibited a monofractal behavior (narrow spectrum) with scaling exponents
close to uncorrelated dynamics. We also remark the fact that for shuffled data, scaling exponents are close to those observed in original data, whereas phase randomization resulted in uncorrelated dynamics, suggesting that the observed correlation
exponents for the original simulated data are mainly related to the distribution of

Fig. 5.15 Results of multi-fractal analysis for shuffled and phase randomized heat release sequences. (**a**) Width of the singularity spectrum $\Delta h = h_{\max} - h_{\min}$ versus ϕ. Phase randomized data are described by a narrow spectrum even for intermediate values of ϕ, indicating a low degree of multifractality. For shuffled data, the width of the spectrum is quite similar to that observed in original data. (**b**) Dominant local Hurst exponent h^* versus ϕ. We observe that for phase randomized data, the fluctuations are dominated by uncorrelated dynamics whereas for shuffled data, the dominant dynamics is similar to that observed in original data (Fig. 5.14)

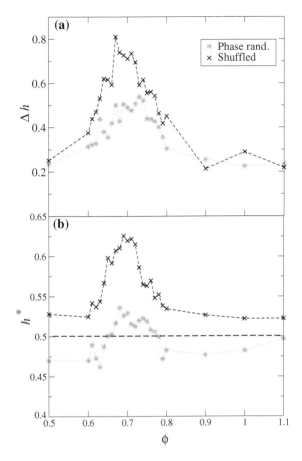

events. More specifically, multifractality and correlated behavior could be associated to the characteristics of the probability density function, this is supported by the fact that shuffled data almost do not show changes in the scaling exponents with respect to the original data [30]. According to the statistics presented in previous studies for both experimental and simulated data [6, 21], the kurtosis and skewness are quantities which dramatically increase for intermediate values of the fuel–air ratio. All these results support the argument that for intermediate values of ϕ, mainly around $\phi \simeq 0.67$, a kind of dynamical transition occurs corresponding to the so-called *engine stability limit*, which according to Heywood is found around $\phi \simeq 0.65$ [20]. This limit separates the regime where the combustion is still complete in all cycles and the regime of partial burn, which for very lean conditions lead to many misfiring events.

5.3.4 Network Analysis

In order to get further insight, we explore additional organizational features of heat release fluctuations by means of *network analysis* [31, 32]. The key step to perform this analysis is to map the times series into a network, allowing to observe correlation properties which complement the framework of dynamical characterization of cycle-to-cycle fluctuations. We use the *visibility graph method* to construct the heat release network [33, 34]. Briefly we explain the method: given the time series, nodes are the values, and two-point values are linked if there is a straight line that connects the series data, with the restriction that this line does not intersect any intermediate data height. According to this, any two time series values with coordinates (t_m, q_m) and (t_n, q_n) are connected if the following criteria are satisfied for any intermediate data,

$$q_i < q_n + (q_m - q_n)\frac{t_m - t_i}{t_n - t_m} \tag{5.4}$$

where (t_i, q_i) are the coordinates of intermediate data.

One advantage of this method is that is invariant under several transformations of the time series such as translation, both horizontal and vertical rescaling and linear trends. One important result from the visibility graph method is that the structure of the signal is also characterized by means of network theory. For instance, the *degree distribution*, $P(k)$, which represents the number of nodes (values) with k links, obtained from uncorrelated series with uniform distribution results in an exponential distribution, whereas periodic time series lead to regular graphs with a peak-shaped distribution of the connectivities. For fractal time series, the emerging graph is characterized by a scale-free degree distribution, where the exponent α characterizes the connectivities.

We proceed to construct the network from the heat release sequences for different values of ϕ. In Fig. 5.16, we show representative networks obtained from short sequences of data. This visual representation of the visibility networks shows representative networks for different values of ϕ. Here, the size of the nodes is according to the degree of each node, that is, highly connected nodes are bigger than nodes with low connectivities. A color scale is also used to indicate the connectivities, dark blue indicates low connectivity, while red color represents high degree. We observe that most of the nodes have low connectivity, while a few are highly connected. This feature is observed especially for intermediate fuel ratio values ($\phi \simeq 0.65$).

Next, we construct the *cumulative degree distribution* to characterize the connectivities for several values of ϕ. To do this, we calculate,

$$G(k) = \int_k^\infty p(r)dr \tag{5.5}$$

where $p(r)$ is the probability density function of cycles with heat release between r and $r + dr$. We analyze the form of $G(k)$ for several values of ϕ. The results of

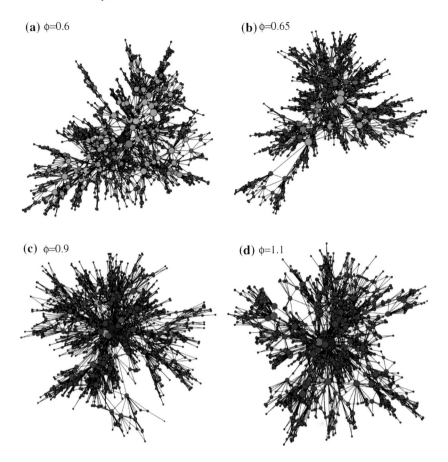

(a) φ=0.6
(b) φ=0.65
(c) φ=0.9
(d) φ=1.1

Fig. 5.16 Representative networks emerging from the visibility graph method for different values of the fuel ratio φ

these calculations are presented in Fig. 5.17. We find that $G(k)$ follows a power law behavior with slight variations as ϕ changes. This plot indicates that for values of connectivity between 10 and 100, the data from all the values of ϕ are consistent with a power law decay. The observed exponent $(G(k) \sim k^{\alpha-1})$, $\alpha - 1 = 1.5$, is related to Brownian-type fluctuations, which results from the integration of white noise increments. Specifically, for low values of connectivity $10 \leq k \leq 10^2$, the probability distribution follows a power law behavior for all the values of ϕ. As the degree increases, for low and high values of ϕ, the probability is smaller than the probability observed at intermediate values, indicating that large connected nodes are more likely to occur for fuel–air ratios around $\phi \simeq 0.65$. From a dynamic energetic point of view, for intermediate fuel–air ratios the presence of large/small heat release values is to a great extent reduced, which permits that one heat release event can '*observe*' other point values located beyond the nearest neighbor.

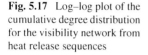

Fig. 5.17 Log–log plot of the cumulative degree distribution for the visibility network from heat release sequences

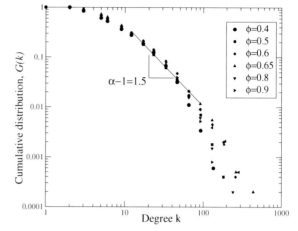

5.3.5 Wavelet Analysis

Wavelet-based techniques are being increasingly used for time series analysis in a variety of applications [35]. They are particularly useful for the analysis of transient and intermittent processes. Wavelet analysis may be performed using a *continuous wavelet transform* (CWT) or a *discrete wavelet transform* (DWT). We will use a CWT to analyze the cyclic variability in the heat release time series obtained by Curto-Risso et al. [21, 36] using quasi-dimensional model simulations of a spark ignition engine fueled by gasoline and gasoline-ethanol blends. The CWT maps the spectral characteristics of a time series on to a time-frequency (time period) plane from which the various periodicities and their temporal variations, if any, can be discerned by visual inspection. In their recent work, Sen et al. have applied both CWT [37–40] and DWT [41] to analyze experimental data from internal combustion engines and investigated the cycle-to-cycle variations in the indicated mean effective pressure (IMEP) and heat release.

The analysis methodology of CWT has been described in detail by Torrence and Compo [42]. We summarize the main steps below. A mother wavelet is chosen and its convolution with the time series signal is computed. This convolution is defined as the CWT of the time series. The CWT is computed by manipulating the mother wavelet over the time series in two ways: it is moved to various locations on the time series, and it is stretched or squeezed. If the wavelet locally matches the shape of the time series, then a large transform value is obtained. If, on the other hand, the wavelet and the time series do not correlate well, a low value of the transform will result. The squared modulus of the CWT, representing the signal energy, is defined as the *wavelet power spectrum* (WPS). The WPS is plotted on a time-frequency (or time-period) plane, which depicts the various periodicities of the time series and their temporal variations. For the heat release time series considered here, the WPS is plotted on a plane with the cycle number and period (cycle) as the two axes. From the WPS, another useful quantity called the *global wavelet spectrum* (GWS) can be computed. The GWS is the average of the WPS over all time, and is analogous to a smoothed Fourier spectrum. From the locations of the peaks in the GWS, the dominant periodicities of the time series can be identified. The Morlet wavelet has been used as a mother wavelet for the analysis of periodic and quasi-periodic signals in a variety of applications. It contains a center frequency (also called order parameter) which controls the frequency resolution of the CWT. In our analysis, we used a Morlet wavelet of order six as the mother wavelet. This choice provides a good balance between the time and frequency resolutions. The interested reader is referred to [42] for further methodological details.

First, we apply a CWT to the heat release time series of Curto-Risso et al. [21] for a gasoline engine (with no ethanol added), and investigate the effect of changing the fuel-air ratio on the cycle-to-cycle variations. The fuel–air ratio is changed from $\phi = 0.7$ and 0.9 for lean mixtures, to $\phi = 1.0$ for a stoichiometric mixture. Figure 5.18 illustrates the WPS and GWS of the various time series. In the WPS, the colors red and blue represent high and low power, respectively, with the other colors designating intermediate power levels. The area below the U-shaped curve represents the cone of influence (COI). For the computation of the WPS, the time series is padded with zeros at both ends. The zero padding leads to edge effects which make the WPS inside the COI unreliable; the WPS inside the COI should therefore be used with caution [42]. Consider the WPS shown in the top panel of Fig. 5.18, which applies to $\phi = 0.7$. This figure shows the presence of both high-frequency (low period) intermittent fluctuations and more persistent low-frequency variations in the heat release time series. Intermittency is characterized by sudden bursts of power separated by intervals of very low power or almost quiescent intervals. The low-frequency variations are most dominant around the 56-cycle period and persist over several engine cycles. The dominant periodicities with corresponding peaks can be discerned from the GWS shown on the right.

Next we consider the results for $\phi = 0.9$. The WPS and GWS for this case are depicted in the middle panel of Fig. 5.18. As in the WPS for $\phi = 0.7$ (upper panel), this WPS also shows high-frequency intermittent fluctuations and low-frequency variations, but a comparison of the GWS in this panel and the top panel reveals that the overall spectral power is significantly reduced. The WPS and GWS for a stoichiometric mixture ($\phi = 1.0$) are presented in the bottom panel in Fig. 5.18. Both high-frequency intermittency and low-frequency periodicities are also seen in this WPS, and the GWS in this figure indicates an additional reduction in the overall spectral power implying that the cycle-to-cycle heat release variations are further diminished.

We now examine the cyclic variability of heat release in a spark-ignition engine fueled by gasoline-ethanol blends. We will use the model simulations of Curto-Risso et al. [36]. We consider that ethanol is added in the proportions of 5, 10, and 25 % by volume, and investigate the effect of ethanol addition on the cycle-to-cycle variations of heat release for fixed fuel–air ratios: $\phi = 0.7$, 0.9, and 1.0, which were used for the gasoline engine with no ethanol added. Figure 5.19 depicts the WPS and GWS of the heat release time series for the lean mixture with $\phi = 0.7$, and ethanol volume

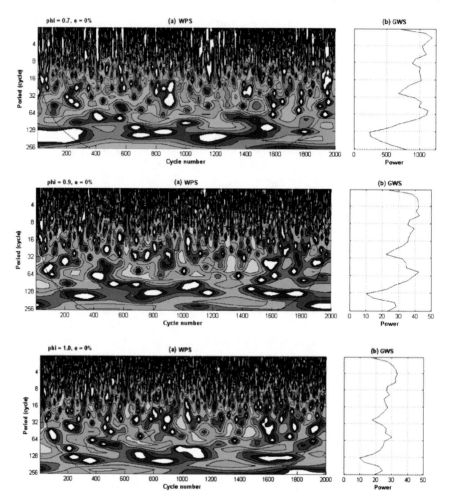

Fig. 5.18 Wavelet power spectrum (WPS) and global wavelet spectrum (GWS) of the heat release time series of the simulated spark ignition engine fueled by gasoline (no ethanol added), with equivalence ratio of 0.7, 0.9, and 1.0

fractions of 5, 10, and 25 %. As in the case of the gasoline engine with no ethanol addition, the WPS in each of the three panels in this figure shows the presence of both high-frequency (low period) intermittent fluctuations and more persistent low-frequency variations. Furthermore, the GWS plots in the three panels indicate that the overall spectral power progressively decreases with increase in ethanol content. Figures 5.20 and 5.21 present the WPS and GWS for $\phi = 0.9$, and 1.0, respectively. The WPS plots in this figure show the occurrence of high-frequency intermittent fluctuations; however, the low-frequency variations are less persistent and rather intermittent in nature. In addition, from the GWS in each of the Figs. 5.20 and 5.21,

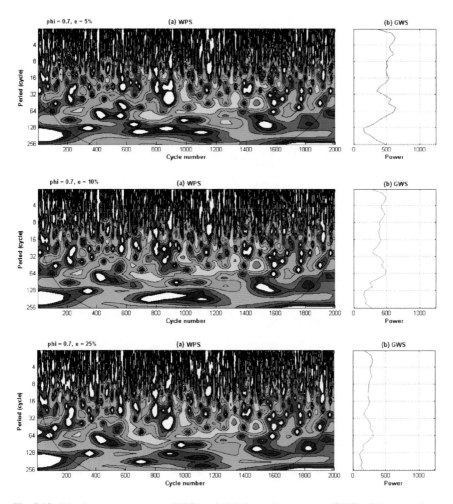

Fig. 5.19 Wavelet power spectrum (WPS) and global wavelet spectrum (GWS) of the heat release time series of the simulated spark ignition engine fueled by gasoline-ethanol blends, with equivalence ratio of 0.7, and ethanol volume fractions 5, 10, and 25 %

we see that the overall spectral power tends to decrease with an increase in ethanol content.

In summary, the results of wavelet analysis reveal that the cyclic heat release variations in a spark ignition engine fueled by gasoline an gasoline-ethanol blends exhibit multi-scale dynamics consisting of high-frequency intermittent fluctuations and low-frequency oscillations. For a gasoline engine (with no ethanol added), the cyclic heat release variations are reduced as the composition of the combustible mixture goes from fuel-lean toward stoichiometric. In addition, for a fixed equivalence ratio, the cyclic heat release variations can be reduced by blending gasoline with ethanol.

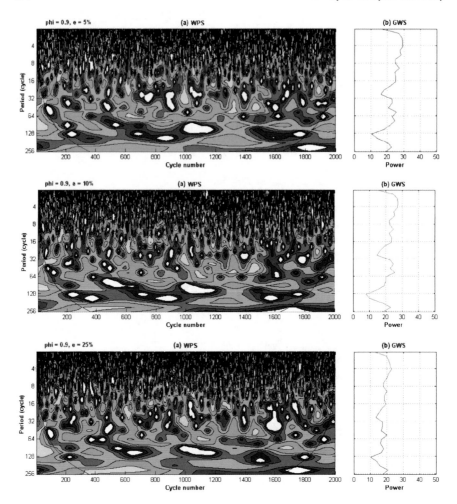

Fig. 5.20 Wavelet power spectrum (WPS) and global wavelet spectrum (GWS) of the heat release time series of the simulated spark ignition engine fueled by gasoline-ethanol blends, with equivalence ratio of 0.9, and ethanol volume fractions 5, 10, and 25 %

5.4 Energetic Properties with Variability

The main goal of this section is to show a probabilistic interpretation of cyclic variability in reference with the most important energetic indicators for a spark ignition engine, assuming that quasi-dimensional simulations with stochastic variations in a combustion parameter such as l_t are capable to reproduce the main experimental features of cyclic variability. As we described in Sect. 5.2, representative cases of the evolution in time of heat release, Q_r, are presented in Fig. 5.2 for several values of the fuel–air ratio, ϕ. We recall that a noisy behavior is observed due to the introduction

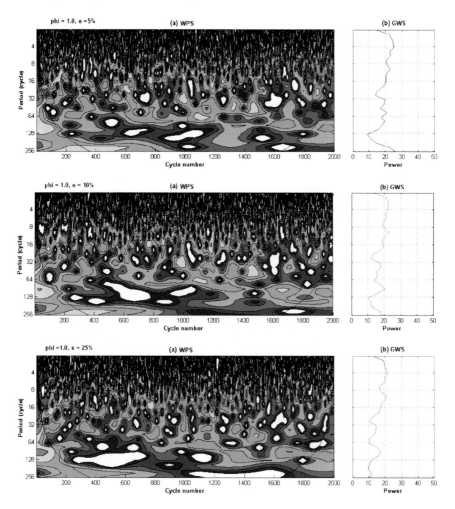

Fig. 5.21 Wavelet power spectrum (WPS) and global wavelet spectrum (GWS) of the heat release time series of the simulated spark ignition engine fueled by gasoline-ethanol blends, with equivalence ratio of 1.0, and ethanol volume fractions 5, 10, and 25 %

of a stochastic term in the simulations, with a decreasing amplitude as the fuel–air ratio is close to unity. Specially, at low ϕ is clear that poor heat release cycles (under the mean value) are frequent. We shall quantitatively analyze this fact hereinafter. This behavior is qualitatively similar to those reported for real data [4] and agrees with previous results discussed at the beginning of this chapter about the complexity of the heat release fluctuations by means of correlation dimension, monofractal, and multifractal analyses [43]. We have shown that over short scales, these fluctuations are characterized by the presence of short-term correlations while for large scales they resemble uncorrelated white noise.

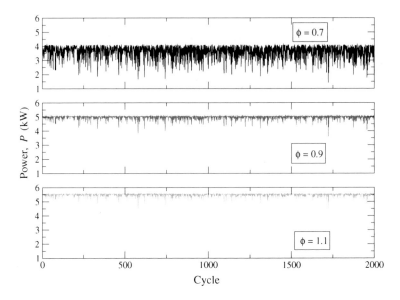

Fig. 5.22 Cycle-to-cycle evolution of the power output, P, for several values of the fuel–air ratio, ϕ

The evolution in time of power and efficiency is presented in Figs. 5.22 and 5.23, respectively, for the same three values of the fuel–air ratio. It is worth mentioning that η_f is not a thermodynamic efficiency (and thus a priori is independent of Q_r), but a fuel conversion efficiency as defined in [44] (see Sect. 3.1.1.2), $\eta_f = P/(\dot{m}_f \, Q_{\text{LHV}})$, where $\dot{m}_f$ is the mass of fuel entering in the cylinder each cycle divided by the cycle duration and Q_{LHV} is the lower heating value of the fuel. As before for heat release, we note a noisy behavior both in P and η_f with a decreasing amplitude as the fuel–air ratio increases.

In order to analyze and compare these time series of energetic quantities, we calculate some usual statistical parameters: the average value (μ), the standard deviation, (STD), the coefficient of variation, (COV), the skewness S, and the kurtosis, K, defined in Sect. 5.2. In Fig. 5.24, we show the results of the calculations of these statistical parameters for several fuel–air ratio values. We are not aware of a simultaneous analysis of P, Q_r and η_f in the literature, neither experimental nor theoretical. The mean value, μ, monotonically increases as the fuel–air ratio increases for P and Q_r whereas for η_f, it shows a convex behavior with a maximum value around the fuel–air ratio $\phi^* \simeq 0.85$. For the standard deviation, σ, the three variables show a decreasing behavior as the fuel–air ratio increases. The coefficient of variation, COV, i.e., the ratio between the dispersion and the mean, also monotonically decreases with the fuel–air ratio for all the time series, but numerically is quite different depending on the considered time series. For the power and the efficiency it reaches 0.25 at low fuel–air ratios. It is 2.5 times larger than for the heat release. The kurtosis, K, for P and η_f, increases until reaching a saturation value around $\phi \simeq 0.9$, while the

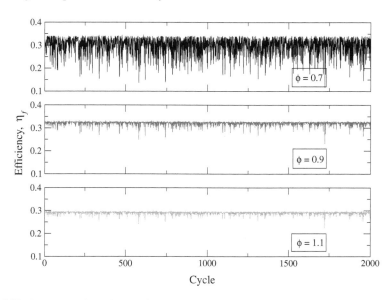

Fig. 5.23 Cycle-to-cycle evolution of the fuel conversion efficiency, η_f, for several values of the fuel–air ratio, ϕ

corresponding K for Q_r is characterized by a peaked convex function with a maximum value located in $\phi = 0.8$. The skewness, S, shows a behavior opposed to that of K for each energetic function. It is almost always negative, indicating (as shown in Figs. 5.2, 5.22, and 5.23) that poor combustion events are abundant. For heat release this is especially intense at fuel–air ratios between $\phi = 0.7$ and 0.9. For stoichiometric and over stoichiometric mixtures, S for Q_r becomes slightly positive indicating the existence of high heat release combustion cycles. The same kind of evolution for the statistical parameters of heat release time series was recently experimentally found by Sen et al. [6] for a real Ford 4.6 L V8 spark ignited engine working at 1,200 rpm as commented in Sect. 5.2.

At sight of Fig. 5.24, it is worth mentioning that the evolution of the statistical properties with the fuel–air ratio for power output and efficiency is very similar (except for the mean value). As we consider a fuel conversion efficiency, as mentioned above, this means that the denominator in η_f (a constant multiplied by the mass of fuel inside the cylinder in each cycle), which makes the difference between η_f and P, has not a definitive influence on the evolution with the fuel-air ratio of σ, K, and S. In principle, $\dot{m}_f$ is fluctuating from one cycle to the following because of cyclic variability, that slightly changes the thermodynamic conditions and residual gases at exhaust, thus affecting the next intake process. But our results show that the fluctuations of $\dot{m}_f$ do not strongly affect the dependence of σ, K, and S with the fuel-air ratio for η_f and P. This is because our model considers that the effect of small variations in the fuel mass (due to changes in the thermodynamic state of residual gases) is overshadowed by the main cause of cyclic variability, that is the

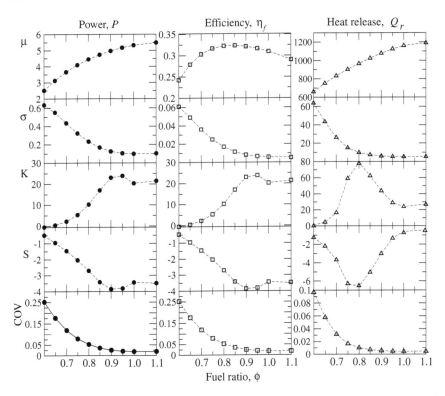

Fig. 5.24 Statistical parameters associated to power output, efficiency, and heat release for several values of the fuel–air ratio, ϕ. Mean value (μ. Units are kW for power and J for heat release), standard deviation (σ), coefficient of variation (COV), skewness (S), and kurtosis (K) are shown. In all the plots we have used *solid* or *dashed lines* to connect the symbols as a guide to the eye

turbulent combustion process. With respect to the evolution with ϕ of the mean value of power output and efficiency time series, power output monotonically increases as the mixture in the combustion chamber is more fuel rich, and of course the same happens for $\dot{m}_f$ in such a way that its ratio first increases up to approximately $\phi = 0.8$ or 0.85, and afterwards decreases as the mixture approaches stoichiometry.

Figure 5.25 shows the results of the fractal dimension at three different values of ϕ. For both performance output functions (η_f and P), the scaling behavior is described by a single exponent $D \simeq 1.5$, which is an indicative of uncorrelated white noise fluctuations. In contrast, for heat release, Q_r, the scaling behavior reveals two regimes (for $0.6 < \phi < 0.8$), over short scales the fractal dimension is close to $D_S \simeq 1.35$, while for large ones $D_L \simeq 1.5$ [43]. These results reveal that while Q_r exhibit short-term correlations for intermediate values of ϕ, fluctuations in P and η_f are processes with no memory (white noise). A more detailed analysis of the crossover between D_S and D_L observed in Q_r for the interval $0.6 < \phi < 0.8$ can be found in [43].

Fig. 5.25 Plot of log $\langle L(k) \rangle$ versus $\log k$ of heat release Q_r, power P and efficiency η_f, for several values of the fuel–air ratio ϕ

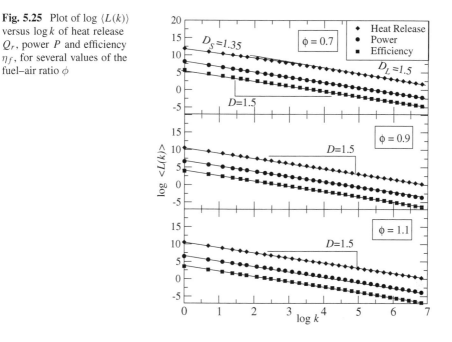

5.4.1 Correlations Between Heat Release and Engine Performance

In order to further explore the effect of the fuel–air ratio on the behavior of the performance output functions, we construct scatter plots of P and η_f as functions of the heat release. Figures 5.26 and 5.27 show the scatter plots of P versus Q_r and η_f versus Q_r, revealing that the relationship between performance functions and the heat release is far to be linear. For lean mixtures (fuel–air ratios under approximately 0.8), the highest Q_r lead to the highest power output or efficiency. But for $\phi > 0.8$, a convex-shaped behavior is observed in both cases. One can roughly identify the values of Q_r^* which lead to the maximum value in P and η_f for each fuel–air ratio. For $Q_r > Q_r^*$, we observe that they lead to values in P smaller than the maximum. The same is observed for the plot of η_f versus Q_r. In fact, the right wing of the dots distribution observed for $\phi > 0.8$, indicates that even when the heat release is high, the corresponding performance output values are smaller than the maximum which is reached at a smaller Q_r value.

We notice that for low fuel–air ratios ($\phi < 0.7$), two clouds of dots (P and η_f) can be roughly identified for low values of Q_r. In both Figs. 5.26 and 5.27, some features suggesting bifurcations appear. However, in real engines, seemingly period doubling bifurcations have not been found [4]. In any case, for the usual values of ϕ for real engines, the zone of bifurcations is outside from typical optima values of power and efficiency, which typically are within the region that represents a good compromise between both high power and high efficiency [45–47].

Fig. 5.26 Scatter plot of the
power versus the heat release
for several values of the fuel–
air ratio, ϕ

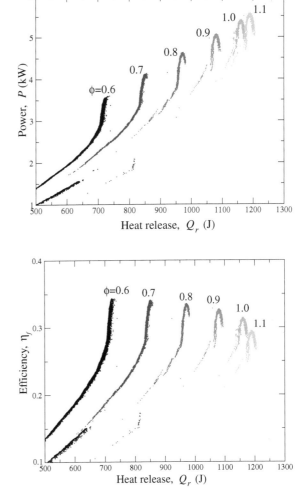

Fig. 5.27 Scatter plot of the
efficiency versus the heat
release for several values of
the fuel–air ratio, ϕ

We also explore the behavior of power-efficiency plots to evaluate the effect of
stochastic variations together with the fuel–air ratio on these performance output
functions. In Fig. 5.28, we show the plot of P versus η_f for several fuel–air ratio
values. It was obtained by considering both P and η_f as fluctuating functions of the
fuel ratio, ϕ, taken as a parametric variable. This figure is the probabilistic counterpart
of the power-efficiency curves that are usual in thermodynamic optimization for
both internal [47–51] and external combustion [52–55] models.[3] This plot supports
a clear way to obtain the maximum achievable efficiency for a certain fixed power
requirement under noisy conditions.

---

[3] See Fig. 3.20 in Chap. 3 and the $P - \eta_f$ curves used with optimization purposes in Chap. 4.

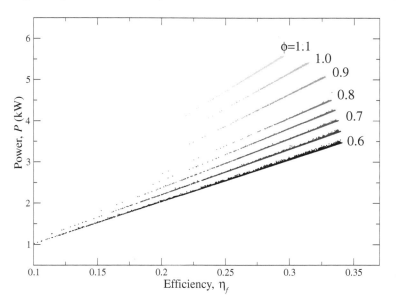

Fig. 5.28 Scatter plot of the power versus the efficiency for several values of the fuel–air ratio ϕ

As shown in Fig. 5.24, for the lowest equivalence ratios the distributions for both power and efficiency are quite flat, less peaked than a Gaussian (K is lower than 3) and present an asymmetric left tail (S is negative). Opposite, at high equivalence ratios the observed variations of both power and efficiency are more peaked around the mean than a Gaussian, while the asymmetric left tails still survive. This fact is consistent with the results from the fractal analysis. The influence of the other two statistical parameters, mean and standard deviation, is better reflected in Fig. 5.29, where we show the mean value of $P = P(\eta_f)$ that would represent the usual power-efficiency curve together with the standard deviation (as a gross measure of the existence of irreversibilities during engine evolution) both of the power output and of the efficiency. From the figure, the following points are remarkable: (a) the monotonic decrease of standard deviation as fuel–air ratio is close to unity for both power and efficiency; (b) the monotonic increase of the mean value for power accomplished by a convex behavior of efficiency (note the maximum of its mean value at intermediate fuel–air ratio, see Fig. 5.24). A clear consequence is the observed loop-shaped power versus efficiency plots, which clearly reflect in a probabilistic manner the well-known features of the noncoincident maximum power and maximum efficiency points. Furthermore, as usual for stationary engines, the optimal values for power and efficiency are in between these points.

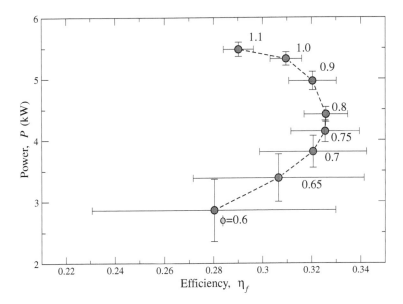

Fig. 5.29 Power-efficiency plots showing the mean value and the standard deviation in both axis for different values of ϕ

5.5 Concluding Remarks on Variability

Finally, we summarize the results discussed in this chapter. We have developed a simulation scheme in order to reproduce the experimentally observed fluctuations of heat release in an Otto engine. As we discussed in the previous chapters, this model relies on the first principle of thermodynamics applied to open systems that allows to build up a set of first-order differential equations for pressure and temperature inside the cylinder. Combustion requires a particular turbulent model giving the evolution of the masses of unburned and burned gases in the chamber as a function of time. We consider a model where inside the approximately spherical flame front there are unburned eddies of typical length l_t. The model incorporates two other parameters, the position of the kernel of combustion, R_c, and a convective characteristic velocity u_t (related with turbulence), that a priori could be important in order to reproduce the observed cycle-to-cycle variations of heat release. Actually, we have checked the influence of each of those parameters when they are considered as stochastic. It is possible from experimental results in the literature to build up log-normal probability distributions for l_t and u_t, and gaussian for R_c. The results of our simulations with these stochastic distributions show that the consideration of l_t or u_t as stochastic is essential to reproduce experiments. These parameters can be considered as independent or related through an empirical correlation. Both possibilities lead to similar results. On the contrary, a stochastic component on R_c only provokes noisy points in the return maps for any fuel ratio. In the second part of our analysis, the heat release sequences were analyzed from the fractal-time organization perspective. Our results showed that there are important changes in the dynamical behavior as ϕ changes, particularly, transitions from a weak anti-correlated to a positive correlated behavior for short scales.

Moreover, multifractal analysis revealed that for intermediate values of ϕ, the singularity spectrum is broad, confirming the multifractality of the signal. In contrast, for high and low fuel–air ratio values, an almost linear multifractal spectrum was identified. The statistics of the singularity spectrum further revealed that for intermediate values of ϕ, the local Hurst exponents cover a broad range related to multifractality. Moreover, visibility network analysis confirmed the complexity of heat release signals for intermediate values of the fuel ratio, that is, high connectivities in a few nodes were observed at intermediate ϕ values, whereas for high and low values most of the nodes have low degree.

From the energetic analysis, we can conclude that for intermediate and near stoichiometric fuel–air ratios, both power and efficiency display a maximum in terms of heat release, i.e., during the fluctuating evolution of heat release, the highest value do not guarantee a maximum power output or efficiency. As fuel–air ratio tends to unity, dispersion (as a gross measure of cyclic variability) in both axis (power and efficiency) diminishes. The mean value of efficiency and power for each fuel–air ratio recovers the well-known power versus efficiency curves widely used in thermodynamic optimization. Moreover, the energetic analysis presented is consistent with typical energetic results of steady-state studies represented by average values only. In fact, low values of fuel–air ratio ($\phi < 0.7$) are concomitant with large fluctuations of power and efficiency. This result is in agreement with the fractal and multifractal studies of the same system. On the other hand, for large values of fuel–air ratio ($\phi > 0.9$), we observe regimes of performance with reduced top values for efficiency (considering that efficiency is a cloud of dots, Fig. 5.27) and high power characterized by small fluctuations (Fig. 5.29). The maximum efficiency point is located in between these two regimes. We finish by recalling attention about the importance of quasi-dimensional models in order to get performance data for heat engines which can be used to elaborate thermodynamic models which, in turn, offer a reliable comparison with observed indicators.

References

1. J. Keck, in *Proceedings of Nineteenth Symposium (International) on Combustion* (The Combustion Institute, Pittsburgh, 1982), pp. 1451–1466
2. G. Beretta, M. Rashidi, J. Keck, Combust. Flame **52**, 217 (1983)
3. J. Keck, J. Heywood, G. Noske, SAE Paper 870164 (1987)
4. C.S. Daw, M.B. Kennel, C.E.A. Finney, F.T. Connolly, Phys. Rev. E **57**, 2811 (1998)
5. G. Litak, T. Kaminski, J. Czarnigowski, D. Zukowski, M. Wendeker, Meccanica **42**, 423 (2007)
6. A. Sen, G. Litak, C. Finney, C. Daw, R. Wagner, Appl. Energ. **87**, 1736 (2010)
7. J.B.J. Green, C.S. Daw, J.S. Armfield, R.M. Wagner, J.A. Drallmeier, M.B. Kennel, P. Durbetaki, SAE Paper 1999-01-0221 (1999)

8. C.S. Daw, C.E.A. Finney, J.B. Green, M.B. Kennel, J.F. Thomas, F.T. Connolly, SAE Paper 962086 (1996)
9. E. Abdi Aghdam, A.A. Burluka, T. Hattrell, K. Liu, G.W. Sheppard, J. Neumeister, N. Crundwell, SAE Paper 2007-01-0939 (2007)
10. C. Stone, A. Brown, P. Beckwith, SAE Paper 960613 (1996)
11. P. Grassberger, I. Procaccia, Phys. Rev. Lett. **50**, 346 (1983)
12. P. Grassberger, I. Procaccia, Physica D **9**, 189 (1983)
13. H. Kantz, T. Schreiber, *Non-Linear Time Serie Analysis* (Cambridge University Press, Cambridge, 2004)
14. R. Hegger, H. Kantz, T. Schreiber, Chaos **9**, 413 (1999)
15. C.K. Peng, S. Havlin, H.E. Stanley, A.L. Goldberger, Phys. Rev. Lett. **70**, 1343 (1995)
16. G. Rangarajan, M. Ding, Phys. Rev. E **61**(5), 4991 (2000). doi:10.1103/PhysRevE.61.4991
17. T. Higuchi, Physica D **46**(2), 254 (1990). doi:10.1016/0167-2789(90)90039-R
18. G. Gálvez-Coyt, A. Munoz-Diosdado, J.L. del Río-Correa, F. Angulo-Brown, Fractals **18**(2), 235 (2009)
19. D.T. Schmitt, P.C. Ivanov, Am. J. Physiol. **293**(5), R1923 (2007). doi:10.1152/ajpregu.00372. 2007
20. J. Heywood, *Internal Combustion Engine Fundamentals* (McGraw-Hill, New York, 1988), Chap. 4, pp. 100–154
21. P.L. Curto-Risso, A. Medina, A. Calvo Hernández, L. Guzmán-Vargas, F. Angulo-Brown, Appl. Energ. **88**, 1557 (2011)
22. D. Scholl, S. Russ, SAE Paper 1999-01-3513 (1999)
23. J. Theiler, S. Eubank, A. Longtin, B. Galdrikian, J. Doyne Farmer, Physica D **58**, 77 (1992)
24. T. Schreiber, A. Schmitz, Physica D **142**(3–4), 346 (2000)
25. J. Feder, *Fractals* (Plenum Press, New York, 1988)
26. B.B. Mandelbrot, *The Fractal Geometry of Nature*, 2nd edn. (Freeman, San Francisco, 1982)
27. J.F. Muzy, E. Bacry, A. Arneodo, Phys. Rev. Lett. **67**(25), 3515 (1991)
28. P.C. Ivanov, L. Amaral, A. Goldberger, S. Havlin, M. Rosenblum, Z. Stuzik, H. Stanley, Nature **399**, 461 (1999)
29. A.L. Goldberger, L.A.N. Amaral, L. Glass, J.M. Hausdorff, P.C. Ivanov, R.G. Mark, J.E. Mietus, G.B. Moody, C.K. Peng, H.E. Stanley, Circulation **101**(23), e215 (2000 (June 13)). http://circ.ahajournals.org/cgi/content/full/101/23/e215
30. J.W. Kantelhardt, S.A. Zschiegner, E. Koscielny-Bunde, A. Bunde, S. Havlin, H.E. Stanley, Multifractal detrended fluctuation analysis of nonstationary time series (2002). http://arxiv. org/abs/physics/0202070
31. D. Watts, S. Strogatz, Nature **393**, 409 (1998)
32. M. Newman, SIAM Rev. **45**, 167 (2003). http://epubs.siam.org/doi/abs/10.1137/ S003614450342480
33. L. Lacasa, B. Luque, F. Ballesteros, J. Luque, J.C. Nun, Proc. Natl. Acad. Sci. U S A **105**, 4972 (2008)
34. L. Lacasa, B. Luque, J. Luque, J.C. Nun, Europhys. Lett. **86**, 30001 (2009)
35. P. Addison, *The Illustrated Wavelet Transform Handbook* (Institute of Physics Publishing, Bristol, 2002)
36. P.L. Curto-Risso, A. Medina, A. Calvo Hernández, in *24th International Conference on Efficiency, Cost, Optimization, Simulation and Environmental Impact* (Novi Sad, Serbia, 2011)
37. A. Sen, J. Zheng, Z. Huang, Appl. Energ. **88**, 2324 (2011)
38. A. Sen, S. Ash, B. Huang, Z. Huang, Appl. Therm. Eng. **31**, 2247 (2011)
39. A. Sen, G. Litak, C. Edwards, C. Finney, C. Daw, R. Wagner, Appl. Energ. **88**, 1648 (2011)
40. M. Ceviz, A. Sen, A. Küleri, I. Öner, Appl. Therm. Eng. **36**, 314 (2012)
41. A. Sen, J. Wang, Z. Huang, Appl. Energ. **88**, 4860 (2011)
42. C. Torrence, G. Compo, Bull. Amer. Meteorol. Soc. **79**, 61 (1998)
43. P.L. Curto-Risso, A. Medina, A. Calvo Hernández, L. Guzmán-Vargas, F. Angulo-Brown, Physica A **389**, 5662 (2010)
44. J. Heywood, *Internal Combustion Engine Fundamentals* (McGraw-Hill, New York, 1988)

45. F. Angulo-Brown, J. Appl. Phys. **69**, 7465 (1991)
46. A. Calvo Hernández, A. Medina, J.M.M. Roco, J.A. White, S. Velasco, Phys. Rev. E. **63**, 037102 (2001)
47. P. Curto-Risso, A. Medina, A. Calvo Hernández, J. Appl. Phys. **105**, 094904 (2009)
48. J. Gordon, M. Huleihil, J. Appl. Phys. **72**, 829 (1992)
49. A. Fischer, K. Hoffmann, J. Non-Equilib. Thermodyn. **29**, 9 (2004)
50. D. Descieux, M. Feidt, Appl. Therm. Eng. **27**, 1457 (2007)
51. P.L. Curto-Risso, A. Medina, A. Calvo Hernández, Appl. Therm. Eng. **31**, 803 (2011)
52. J. Chen, J. Phys. D: Appl. Phys. **27**, 1144 (1994)
53. A. Durmayaz, O.S. Sogut, B. Sahin, H. Yavuz, Prog. Energ. Combust. **30**, 175 (2004)
54. A. Calvo Hernández, A. Medina, J.M.M. Roco, J. Phys. D: Appl. Phys. **28**, 2020 (1995)
55. S. Sánchez-Orgaz, A. Medina, A. Calvo Hernández, Energ. Convers. Manage. **51**, 2134 (2010)

Appendix A
Derivation of the Basic Differential Thermodynamic Equations

The control volume considered to apply the first principle of thermodynamics is delimited by the cylinder inner walls. During intake and exhaust strokes the thermodynamic system is open and, on the contrary, during compression and expansion it is a closed system, so the different terms appearing in the mathematical expression for the first principle depend on the regarded stroke. We obtain in this appendix, the thermodynamic general differential equations for any stroke considering the system as open in the most widespread case. In the main text, (Sect. 2.2) they are particularized for any stroke during the engine evolution. The differential equations that will be obtained are suitable to be solved in an iterative way through an adequate numerical algorithm.

A.1 Thermodynamic Background

The first principle of thermodynamics for an open system, if potential and kinetic energy variations are considered negligible can be written, in general, as [1]:

$$\dot{E} = \dot{Q} - \dot{W} + \sum_j \dot{m}_j h_j \qquad (A.1)$$

where $\dot{Q}$ is the heat absorbed by the system per unit time and $\dot{W}$ is the work exerted per unit time. $\dot{m}_j$ is the mass rate of each component in the control volume ($\dot{m}_j$ is taken as positive when the mass enters the system). $\sum_j \dot{m}_j h_j$ is the enthalpy change through the boundaries per unit time. The variation with time of the total energy, $\dot{E}$, can be expressed as:

$$\dot{E} = \frac{\mathrm{d}}{\mathrm{d}t}(mh) - \frac{\mathrm{d}}{\mathrm{d}t}(pV) \qquad (A.2)$$

A. Medina et al., *Quasi-Dimensional Simulation of Spark Ignition Engines*,
DOI: 10.1007/978-1-4471-5289-7, © Springer-Verlag London 2014

The *fuel–air ratio*,[1] ϕ, is the ratio between the masses of fuel and air and their stoichiometric counterparts [2]:

$$\phi = \frac{(F/A)}{(F/A)_s} \tag{A.3}$$

If it is assumed that the system is characterized by its temperature, T, pressure p, and equivalence ratio, ϕ, the corresponding state functions depend on three state variables. For instance, the enthalpy is: $h = h(T, p, \phi)$ and its total derivative with time [3]:

$$\dot{h} = \frac{\partial h}{\partial T}\dot{T} + \frac{\partial h}{\partial p}\dot{p} + \frac{\partial h}{\partial \phi}\dot{\phi} \tag{A.4}$$

So from Eq. (A.1) it is obtained:

$$\frac{\mathrm{d}}{\mathrm{d}t}(mh) - \frac{\mathrm{d}}{\mathrm{d}t}(pV) = \dot{Q} - \dot{W} + \sum_j \dot{m}_j h_j \tag{A.5}$$

$$\dot{m}h + m\dot{h} = \dot{Q} - \dot{W} + \sum_j \dot{m}_j h_j + \frac{\mathrm{d}}{\mathrm{d}t}(pV) \tag{A.6}$$

$$m\dot{h} = \dot{Q} - \dot{W} + \sum_j \dot{m}_j h_j - \dot{m}h + \frac{\mathrm{d}}{\mathrm{d}t}(pV) \tag{A.7}$$

Substituting $\dot{h}$ as given by Eq. (A.4) results:

$$\frac{\partial h}{\partial T}\dot{T} + \frac{\partial h}{\partial p}\dot{p} + \frac{\partial h}{\partial \phi}\dot{\phi} = \frac{1}{m}\left(\dot{Q} - \dot{W} + \sum_j \dot{m}_j h_j - \dot{m}h + \frac{\mathrm{d}}{\mathrm{d}t}(pV)\right) \tag{A.8}$$

Taking into account that $c_p = \left(\dfrac{\partial h}{\partial T}\right)_p$, the evolution with time of the temperature is governed by:

$$\dot{T} = \frac{1}{c_p}\left[\frac{1}{m}\left(\dot{Q} - \dot{W} + \sum_j \dot{m}_j h_j - \dot{m}h + \frac{\mathrm{d}}{\mathrm{d}t}(pV)\right) - \frac{\partial h}{\partial p}\dot{p} - \frac{\partial h}{\partial \phi}\dot{\phi}\right] \tag{A.9}$$

If the working fluid is considered as an ideal gas, and assuming that the fuel–air ratio is constant, the partial derivatives of h with respect to p and ϕ can be eliminated in the last equation, and thus:

---

[1] For fuel-lean mixtures $\phi < 1$, for stoichiometric mixtures, $\phi = 1$, and for fuel-rich mixtures, $\phi > 1$.

$$\dot{T} = \frac{1}{mc_p} \left(\dot{Q} - \dot{W} + \sum_j \dot{m}_j h_j - \dot{m} h + \frac{d}{dt}(pV) \right) \tag{A.10}$$

If the time derivative of the total work, $\dot{W}$, is expressed in terms of the volume derivative and simplifying,

$$\dot{T} = \frac{1}{mc_p} \left(\dot{Q} + \sum_j \dot{m}_j h_j - \dot{m} h + V\dot{p} \right) \tag{A.11}$$

It is possible to obtain an equation for the evolution of pressure by differentiating the ideal gas equation of state and substituting the equation for $\dot{T}$ by (A.11) [4]. Both equations are basic to follow the time evolution of the system from a thermodynamic viewpoint. They are valid for any stroke during the system evolution, except during combustion.

A.2 General Formulation for Temperature and Pressure Equations

In this section, we express the thermodynamic evolution equations for pressure and temperature in the particular situation that the system is formed by a gas mixture composed of unburned and burned gases (subscripts u and b respectively), as it happens in the models considered in this monograph. These equations are valid in any step during the system evolution except combustion[2] and are specially useful for numerical resolution within a computer program. $\dot{Q}_u$ and $\dot{Q}_b$ denote heat loss from the unburned or the burned gases. In the processes where do not exist unburned gases inside the cylinder, $m_u = 0$ and then $\dot{Q}_u = 0$, and the same applies for the burned gases. So each term for u or b can appear or not in the equation depending on the stroke. The initial values are given by the external conditions.

- Temperature equation (from Eq. (A.11)) [5, 6].

$$\dot{T} = \frac{1}{\left(m_u c_{p,u} + m_b c_{p,b}\right)} \left[\dot{Q}_u + \dot{Q}_b + \dot{m}_{in} h_{in} - \dot{m}_u h_u + \dot{m}_{ex} h_{ex} - \dot{m}_b h_b + V\dot{p} \right] \tag{A.12}$$

This equation gives the evolution either of the unburned or burned gases in all the strokes except during combustion.

1. During intake (but excluding the overlap period), $\dot{m}_{in} = \dot{m}_u$ and $\dot{m}_{ex} = 0$.
2. During exhaust (but excluding the overlap period), $\dot{m}_{ex} = \dot{m}_b$ and $\dot{m}_{in} = 0$.

---

[2] The equations giving the evolution of temperature and pressure during combustion are those in Sect. 2.2.3.

3. During the overlap period all the terms in Eq. (A.12) can be different from zero, so the equation should be considered in its general form. It will be explained in Appendix B how to evaluate all the mass fluxes. Particularly, the calculation of $\dot{m}_{in}$ requires Eqs. (B.7) and (B.9), and the calculation of $\dot{m}_{ex}$ requires Eqs. (B.18)–(B.21).
4. During compression and expansion, the system is closed and all mass fluxes are null.

- Pressure equation [4].

$$\dot{p} = \left[p \left(\frac{\dot{m}_u}{\rho_u} + \frac{\dot{m}_b}{\rho_b} - \dot{V} \right) \right.$$
$$\left. + \zeta \left(\dot{Q}_u + \dot{Q}_b + \dot{m}_{in}h_{in} - \dot{m}_u h_u + \dot{m}_{ex}h_{ex} - \dot{m}_b h_b \right) \right] \frac{1}{[V(1-\zeta)]} \tag{A.13}$$

with

$$\zeta = \frac{V}{\dfrac{V_u c_{p,u}}{R_u} + \dfrac{V_b c_{p,b}}{R_b}} \tag{A.14}$$

R_u and R_b are the T- and p-independent gas constants for unburned or burned gases, respectively. These equations recover those in Sects. 2.2.1 and 2.2.2 when particularized for the corresponding strokes during the engine evolution. They can be solved through the appropriate numerical algorithms taking as variable either the time or the crankshaft angle, φ, through the relation, $\varphi = \omega t$ where ω is the angular speed. The opening and closing of the intake and exhaust valves are usually fixed at certain values of φ, and the cycle is comprised between $\varphi = 0$ and $\varphi = 4\pi$. So in the crankshaft space these limits are fixed, but in the time space they depend on the angular speed.

References

1. M. Moran, H. Shapiro, *Fundamentals of Engineering Thermodynamics* (Wiley, New York, 2008)
2. J. Heywood, *Internal Combustion Engine Fundamentals* (McGraw-Hill, New York, 1988), Chap. 4, pp. 68–72
3. J. Heywood, *Internal Combustion Engine Fundamentals* (McGraw-Hill, New York, 1988), Chap. 14, pp. 748–819
4. C. Borgnakke, P. Puzinauskas, Y. Xiao, Spark ignition engine simulation models. Tech. rep., Department of Mechanical Engineering and Applied Mechanics. University of Michigan. Report No. UM-MEAM-86-35 (1986)
5. H. Bayraktar, O. Durgun, Energ. Sources **25**, 439 (2003)
6. P.L. Curto-Risso, A. Medina, A. Calvo Hernández, J. Appl. Phys. **104**, 094911 (2008)

Appendix B
Flow Rates and Valve Geometry

B.1 Gas Mass Flow Rates

During intake and exhaust, the thermodynamic system considered is open and it is necessary to quantify the mass rates to solve Eqs. (2.12)–(2.15). Although intake and exhaust flows actually are pulsating, and can be analyzed on a quasi-steady basis. Instead of considering directly a real gas flow, it is usual to assume an equivalent ideal flow as the steady adiabatic frictionless flow of an ideal gas through a restriction with given geometry and dimensions.

The starting point is the first principle for control volumes applied to a compressible flow through a restriction or nozzle. Changes in kinetic energy should be explicitly considered. We express the first principle of thermodynamics as:

$$\dot{E} = \dot{Q} - \dot{W} + \dot{m}_{in}e_{in} - \dot{m}_{out}e_{out} \qquad (B.1)$$

where the term on the left, $\dot{E}$, is the total enthalpy rate, $\dot{Q}$ and $\dot{W}$ respectively the heat added and work done per unit time, $\dot{m}_{in}e_{in}$ the total enthalpy flowing to the control volume, and $\dot{m}_{out}e_{out}$ the total enthalpy flowing out of the control volume. e denotes the total enthalpy per unit of mass: $e = h + c^2/2 + z$ where h is the specific enthalpy, $c^2/2$ the specific kinetic energy, and z the specific potential energy.

Assuming steady flow and rewriting Eq. (B.1) in terms of the initial and final states, and assuming negligible changes on the potential energy[3]:

$$q - w_s = h - h_0 + \frac{c^2}{2} - \frac{c_0^2}{2} \qquad (B.2)$$

where all extensive magnitudes are expressed per unit of mass, and w_s denotes *shaft work*. Introducing the specific heats and assuming that the heat transfer is negligible (the area of the restriction is small) as well as the shaft work:

---

[3] The subscript 0 denotes the state before the restriction (initial state). Variables with no subscript refer to the state after the restriction (final state).

A. Medina et al., *Quasi-Dimensional Simulation of Spark Ignition Engines*,
DOI: 10.1007/978-1-4471-5289-7, © Springer-Verlag London 2014

$$c_p T_0 + \frac{c_0^2}{2} = c_p T + \frac{c^2}{2} \tag{B.3}$$

so the *stagnation enthalpy* ($h_0 = c_p T + c^2/2$) or *stagnation temperature* ($T_0 = T + c^2/(2c_p)$) are constant provided that $c_0 \ll c$. As the process is taken as isentropic:

$$\frac{T}{T_0} = \left(\frac{p}{p_0}\right)^{\frac{\gamma-1}{\gamma}} \tag{B.4}$$

where p_0 is the *stagnation* or *total pressure* ($p_0 = p + \frac{1}{2}\rho c^2$). Introducing the *Mach number* as the ratio between the gas speed and the *sound speed*, $M = c/(\gamma R T)^{1/2}$ and substituting in the definition of T_0:

$$\frac{T_0}{T} = 1 + \frac{\gamma-1}{2}M^2 \tag{B.5}$$

or using Eq. (B.4):

$$\frac{p_0}{p} = \left(1 + \frac{\gamma-1}{2}M^2\right)^{\gamma/(\gamma-1)} \tag{B.6}$$

The ideal gas mass rate flowing through an orifice is $\dot{m}_{\text{ideal}} = \rho A_T c$ where ρ is its density and A_T the transversal area. Using the definition of M and Eq. (B.6):

$$\frac{\dot{m}_{\text{ideal}}\sqrt{\gamma R T_0}}{A_T p_0} = \gamma \left(\frac{p_T}{p_0}\right)^{\frac{1}{\gamma}} \left\{ \frac{2}{\gamma-1} \left[1 - \left(\frac{p_T}{p_0}\right)^{\frac{\gamma-1}{\gamma}} \right] \right\}^{\frac{1}{2}} \tag{B.7}$$

The maximum mass flow for a fixed pair (p_0, T_0) happens when the velocity at the throat or minimum area is the velocity of sound. This flow is named *choked* or *critical* (when the velocity is below velocity of sound the flow is called *subcritical*). In this case, the ratio between the pressure at the throat, p_T and the stagnation pressure, p_0, is given by [1]:

$$\frac{p_T}{p_0} = \left(\frac{2}{\gamma+1}\right)^{\frac{\gamma}{\gamma-1}} \tag{B.8}$$

For pressure ratios under or equal p_T/p_0, the mass flow is maximum:

$$\frac{\dot{m}_{\text{ideal}}\sqrt{\gamma R T_0}}{A_T p_0} = \gamma \left(\frac{2}{\gamma+1}\right)^{\frac{\gamma+1}{2(\gamma-1)}} \tag{B.9}$$

In order to relate the ideal flow to real ones, it introduced the so-called *discharge coefficient*, C_D, as the ratio between the actual mass flow and the ideal one. So Eqs. (B.7) and (B.9) for subcritical and critical conditions are valid for real flows

Table B.1 Parameters required to evaluate the mass flow rates during intake and exhaust depending on the relative pressure inside and outside the cylinder

		p_T	p_0	T_0	A_T	R
$\dot{m}_{in}$	$(p_{in} > p)$	p	p_{in}	T_{in}	$A_{v,in}$	R_{in}
$-\dot{m}_{in}$	$(p > p_{in})$	p_{in}	p	T	$A_{v,in}$	R_{cyl}
$-\dot{m}_{ex}$	$(p > p_{ex})$	p_{ex}	p	T	$A_{v,ex}$	R_{cyl}
$\dot{m}_{ex}$	$(p_{ex} > p)$	p	p_{ex}	T_{ex}	$A_{v,ex}$	R_{cyl}

Table B.2 Basic elements for the geometrical and dynamical description of poppet valves

$L_{v,max}$	Maximum lift
D_v	Head diameter
β	Seat angle
w	Seat width
D_s	Stem diameter
D_p	Inner seat diameter
φ_{op}	Opening angle
φ_{cl}	Closing angle

by substituting, $\dot{m}_{ideal} = \dot{m}_{real}/C_D$. Usually C_D varies in the interval $[0.55, 0.8]$ [2, 3]. In Table B.1, we summarize the parameters required to evaluate mass flow rates during intake and exhaust depending on the relative pressure inside and outside the cylinder.

In Table B.1, $A_{v,in}$ denotes the flux transverse area in the intake valve, $A_{v,ex}$ to the flux transverse area in the exhaust valve, and R_{cyl} to the ideal gas constant of the gas mixture inside the cylinder. With respect to temperatures: T is the temperature inside the cylinder, T_{in} is that in the intake duct, and T_{ex} in the exhaust duct (that could be considered as the average temperature in the exhaust manifold). For pressure, notation is analogous.

B.2 Valve Geometry

In order to calculate the mass flows, it is necessary to know the geometry of the intake and exhaust valves, and so the flow area. Following [4, 5] poppet valves with revolution symmetry are considered. Geometrical definitions are summarized and depicted in Table B.2 and Fig. B.1.

The mass flow through the valves depends on their lift and geometrical characteristics. As valve lifts there are three stages leading to the corresponding flow areas.

(i) $0 < L_v < \dfrac{w}{\sin \beta \cos \beta}$

In this situation, the minimum passage area is a circular cone frustum between the valve and the seat. The flow area in this case is:

Fig. B.1 Intake and exhaust
poppet valves geometry

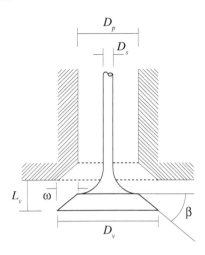

$$A_T = \pi L_v \cos \beta \left[D_v - 2w + \frac{L_v}{2} \sin(2\beta) \right] \tag{B.10}$$

(ii) $\left[\left(\frac{D_p^2 - D_s^2}{4 D_m} \right) - w^2 \right]^{\frac{1}{2}} + w \tan \beta \geq L_v > \dfrac{w}{\sin \beta \cos \beta}$

In this stage, the minimum area has the same form that in the previous one, but
now this surface is not perpendicular to the valve seat [4].

$$A_T = \pi (D_v - \omega) \sqrt{(L_v - w \tan \beta)^2 + w^2} \tag{B.11}$$

(iii) $L_v > \left[\left(\frac{D_p^2 - D_s^2}{4(D_v - \omega)} \right) - w^2 \right]^{\frac{1}{2}} + w \tan \beta$

When the valve lift is large enough, the minimum flow area is no longer between
the valve head and seat, but the port flow area minus the sectional area of the
valve stem. Thus,

$$A_T = \frac{\pi}{4} \left(D_p^2 - D_s^2 \right) \tag{B.12}$$

Proceeding in this way the area is associated to its lift, L_v, but in order to perform
the computations from the dynamical equations it is necessary to relate it with
the crankshaft angle, φ. Usual curves of the valve lift in terms of the camshaft
angle can be found in [4, 6]. A symmetrical timing for a four-stroke engine can
be correctly fitted by this function as a first approximation:

$$L_v = L_{v,\max} \sin^2 \left(\frac{\varphi - \varphi_{\mathrm{op}}}{\varphi_{\mathrm{cl}} - \varphi_{\mathrm{op}}} \pi \right) \tag{B.13}$$

where φ_{op} and φ_{cl} correspond to the opening and closing angles of the valves. The effects of *variable valve timing* for fuel economy improvement were analyzed by Fontana and Galloni [7] by means of CFD simulations.

B.3 Gas Mass Flow Rates and Enthalpy During *Overlapping*

In this section, we analyze how to calculate the mass flow rates and the enthalpies during the overlap period, in which intake and exhaust valves are simultaneously open. Depending on the relative pressures, mass flows of unburned or burned gases go in or out the cylinder.

Denoting by $x_b = m_b/(m_u + m_b)$, the gases flowing into the cylinder through the intake valve are unburned gases and through the exhaust valve flows out a mixture of unburned gases $(1 - x_b)$ and burned gases (x_b). Once the composition of the mixture in the intake and exhaust ducts is known the calculation of enthalpies is straightforward.

For a given equivalence ratio, ϕ, the composition of the intake mixture in terms of its mass fractions is:

$$x_f^* = \frac{\phi r_q}{1 + \phi r_q}; \qquad x_a^* = \frac{1}{1 + \phi r_q} \qquad (B.14)$$

where x refers to mass fractions, the subscripts f and a to fuel and air respectively, and r_q is the ratio between the fuel and air masses in stoichiometric conditions. The asterisk indicates conditions outside the cylinder. So with this notation enthalpies can be obtained in the following way:

(i) The mixture flows into the cylinder through the intake valve:

$$h_{in} = x_f^* h_f + x_a^* h_a \qquad (B.15)$$

(ii) The mixture flows out of the cylinder through the intake valve:

$$h_{in} = (1 - x_b) h_u + x_b h_b \qquad (B.16)$$

(iii) The mixture flows into or out of the cylinder through the exhaust valve (assuming that the gas in the exhaust manifold has the same composition that inside the cylinder):

$$h_{ex} = (1 - x_b) h_u + x_b h_b \qquad (B.17)$$

The mass flow rates can be calculated by means of a mass balance depending on the pressures inside the cylinder and the pressures at intake or exhaust, and the chemical compositions in the intake or exhaust mixtures. The possible situations are summarized in Table B.3.

Table B.3 Possible flows during the overlap period in terms of the relative pressures

Case I	$p < p_{in}$	$p < p_{ex}$	Mixture flows into the cylinder through both valves
Case II	$p < p_{in}$	$p > p_{ex}$	Flow enters through intake and goes out through exhaust
Case III	$p > p_{in}$	$p < p_{ex}$	Flow goes out through intake and enters through exhaust
Case IV	$p > p_{in}$	$p > p_{ex}$	Mixture flows out of the cylinder through both valves

Case I

$$\dot{m}_u = \dot{m}_{in} + (1 - x_b)\,\dot{m}_{ex}$$
$$\dot{m}_b = x_b \dot{m}_{ex} \tag{B.18}$$

Case II

$$\dot{m}_u = \dot{m}_{in} - (1 - x_b)\,\dot{m}_{ex}$$
$$\dot{m}_b = -x_b \dot{m}_{ex} \tag{B.19}$$

Case III

$$\dot{m}_u = -(1 - x_b)\,\dot{m}_{in} + (1 - x_b)\,\dot{m}_{ex}$$
$$\dot{m}_b = -x_b \dot{m}_{in} + x_b \dot{m}_{ex} \tag{B.20}$$

Case IV

$$\dot{m}_u = -(1 - x_b)\,\dot{m}_{in} - (1 - x_b)\,\dot{m}_{ex}$$
$$\dot{m}_b = -x_b \dot{m}_{in} - x_b \dot{m}_{ex} \tag{B.21}$$

In order to improve the precision for the calculation of x_b, it is possible to calculate the burned gases flowing out through the exhaust valve, and then consider that when the sign of the flow in this valve changes, first there is a flow in of burned gases and second a flow in of fresh mixture.

References

1. C. Taylor, *The Internal Combustion Engine in Theory and Practice*, vol. I (The MIT Press, Cambridge, 1994)
2. J. Heywood, *Internal Combustion Engine Fundamentals* (McGraw-Hill, New York, 1988), Chap. App. C, pp. 907–910
3. C.R. Ferguson, *Internal Combustion Engines, Applied Thermosciences* (Wiley, New York, 1986), Chap. 7, pp. 257–263
4. J. Heywood, *Internal Combustion Engine Fundamentals* (McGraw-Hill, New York, 1988), Chap. 6, pp. 220–230
5. C. Taylor, *The internal Combustion Engine in Theory and Practice*, vol. II (The MIT Press, Cambridge, 1994)
6. R. Stone, *Introduction to Internal Combustion Engines* (Macmillan Press, London, 1999), Chap. 6, pp. 276–279
7. G. Fontana, E. Galloni, Appl. Energ. **86**, 96 (2009)

Appendix C
Flame Front Area Calculations

In order to solve the Eqs. (2.35) and (2.36), it is necessary to estimate the flame front area, A_f, during the combustion process. Assuming a spherical flame front, A_f, can be calculated from its radius, R_f, and this estimated by at least two alternative procedures:

- Beretta et al. [1] assume that R_f, is a function of a *characteristic turbulent flame thickness*, L_T, and the *equivalent burned gas radius*, r_b[4] and proposes the following correlation from laboratory measurements:

$$R_f = r_b + L_T \left\{ 1 - \exp\left[-\left(\frac{r_b}{L_T}\right)^2 \right] \right\} \qquad (C.1)$$

where $L_T = u_t \tau_b$ (all the definitions can be found in Sect. 2.3.1) and r_b is calculated from the burned gases volume, V_b, and making use of the approximately spherical flame front hypothesis.

- Bayraktar and Durgun [2] propose to obtain R_f directly from the volume and the spherical flame front approximation. The volume of the gases inside the flame front is calculated from:

$$V_f = V_b + \frac{m_e - m_b}{\rho_u} \qquad (C.2)$$

where V_b is the volume of the burned gases during combustion, $V_b = m_b/\rho_b$. An alternative equation can be used:

$$V_f = V - \frac{m - m_e}{\rho_u} \qquad (C.3)$$

---

[4] The *equivalent burned gas radius* is the radius of an ideal spherical surface, concentric with the flame front, containing all and only burned gases [1].

A. Medina et al., *Quasi-Dimensional Simulation of Spark Ignition Engines*,   157
DOI: 10.1007/978-1-4471-5289-7, © Springer-Verlag London 2014

Here V is the total cylinder volume and m the total gas mass in the cylinder. These are constant quantities so this expression for V_f depends on less varying quantities so the numerical noise is filtered.

A way to start up combustion from simulations it is by taking an arbitrarily small value for V_f (e.g., $V_f = 1\,\text{mm}^3$) [3] or by using any phenomenological correlation as the Wiebe function [Eq. (2.28)] or the cosine burning law [4] to estimate the initial behavior of x_b. Willems and Sierens [5], and Verhelst and Sheppard [6] discuss some procedures to simulate *ignition*. Simplifications are justified when ignition is not critical, for instance to predict indicated work and efficiency for near stoichiometric mixtures at moderate engine speeds. At conditions sensitive to ignition (lean mixtures, high speeds), such simplifications should be carefully analyzed.

In the next sections, different procedures to calculate the flame front area, A_f, from its volume, V_f, (Eq. C.2) are presented, for centered and noncentered spark plug positions.

C.1 Centered Ignition

The four possible configurations for the area of the flame front in the case of centered ignition are shown in Fig. C.1. The usual sequence is: case I, then case II and last case IV. Case III is only possible for very large clearance volumes (that are never achieved in internal combustion engines). First, the radius, R_f, is obtained in terms of the volume, V_f and this in turn allows to calculate the flame area, A_f.

- Case I:

$$V_f = \frac{2}{3}\pi R_f^3 \tag{C.4}$$

$$R_f = \left(\frac{3V_f}{2\pi}\right)^{\frac{1}{3}} \tag{C.5}$$

- Case II:

$$V_f = \frac{\pi}{3}h\left(3R_f^2 - h^2\right) \tag{C.6}$$

$$R_f = \sqrt{\frac{1}{3}\left(\frac{3V_f}{h\pi} + h^2\right)} \tag{C.7}$$

- Case III:

$$V_f = \frac{2\pi}{3}\left[R_f^3 - \frac{1}{8}\left(4R_f^2 - B^2\right)^{\frac{3}{2}}\right] \tag{C.8}$$

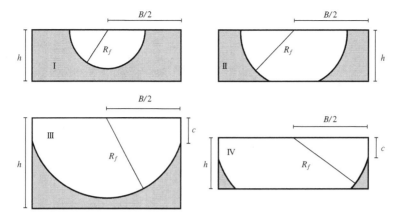

Fig. C.1 Possible flame front configurations for a centered ignition location

- Case IV:

$$V_f = \frac{\pi}{6} \left[\frac{3}{4} B^2 \left(R_f + \sqrt{R_f^2 - \frac{B^2}{4}} \right) - 2 \left(2R_f + h \right) \left(R_f - h \right)^2 \right.$$

$$\left. + \left(R_f + \sqrt{R_f^2 - \frac{B^2}{4}} \right)^3 \right] \tag{C.9}$$

In the cases III and IV, it is necessary to apply any iterative method in order to obtain R_f. The flame front area is obtained from its radius:

- Case I:

$$A_f = 2\pi R_f^2 \tag{C.10}$$

- Case II:

$$A_f = 2\pi R_f h \tag{C.11}$$

- Case III:

$$A_f = 2\pi R_f \left(R_f - \sqrt{R_f^2 - \frac{B^2}{4}} \right) \tag{C.12}$$

- Case IV:

$$A_f = 2\pi R_f \left(h - \sqrt{R_f^2 - \frac{B^2}{4}} \right) \tag{C.13}$$

Fig. C.2 Geometrical definitions for the calculation of the flame front area for a non-centered ignition location

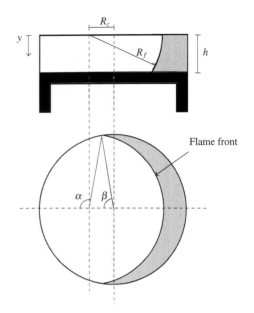

C.2 Noncentered Ignition

C.2.1 Approximate Calculations

There exist in the literature, approximate iterative procedures to estimate the flame front area when $R_c \neq 0$ [7, 8]. In the case the flame front does not reach the piston head, the situation is similar to the cases I and II of the previous section. For the layout drawn in Fig. C.2, Bayraktar [7] proposes the following equations for the area and the volume at the step i ($A_{f,i}$ and $V_{f,i}$), assuming that the flame volume resembles a cylinder:

$$A_{f,i} = \int_0^S 2\,(\pi - \alpha)\,R_{f,i}\,dy \tag{C.14}$$

$$V_{f,i} = \int_0^S \left[\left(\pi - \alpha + \frac{\sin(2\alpha)}{2} \right) \left(R_{f,i}^2 - y^2 \right) + (2\beta - \sin(2\beta))\,\frac{B^2}{8} \right] dy \tag{C.15}$$

where $S = R_{f,i}$ whenever $R_{f,i} < h$ and $S = h$ otherwise. The radius can be calculated from an iterative procedure:

$$R_{f,i+1} = R_{f,i} + \frac{V_f - V_{f,i}}{A_{f,i}} \tag{C.16}$$

Close to convergence ($R_{f,i+1} \simeq R_{f,i}$) and the area is obtained again from Eq. (C.14).

C.2.2 Exact Calculations

In order to calculate V_f within an exact procedure and provided that the flame front is taken as a truncated sphere, seven possible flame front configurations should be discerned during the evolution of the combustion front [9], (Fig. C.3). Denoting by A_c, the area of an horizontal section of the volume of gases inside the flame front at a vertical position y in the cylinder, and by P_c the perimeter of that section:

$$A_c\left(R_f, R_c, B, y\right) = \frac{\beta B^2}{4} + \left(R_f^2 - y^2\right)(\pi - \alpha) - \frac{R_c B}{2}\sin\beta \tag{C.17}$$

$$P_c\left(R_f, R_c, B, y\right) = 2\left(\pi - \alpha\right)\sqrt{R_f^2 - y^2} \tag{C.18}$$

where the angles α and β (see Fig. C.2) are obtained through the following geometrical relations:

$$\alpha\left(R_f, R_c, B, y\right) = \arccos\left[\frac{B^2 - 4\left(R_f^2 - y^2 + R_c^2\right)}{8R_c\sqrt{R_f^2 - y^2}}\right] \tag{C.19}$$

$$\beta\left(R_f, R_c, B, y\right) = \arccos\left[\frac{B^2 - 4\left(R_f^2 - y^2 - R_c^2\right)}{4R_c B}\right] \tag{C.20}$$

In the Fig. C.3, seven possible cases throughout the flame front evolution are specified:

- Case I:

$$V_f = \frac{2}{3}\pi R_f^3 \tag{C.21}$$

$$R_f = \left(\frac{3V_f}{2\pi}\right)^{1/3} \tag{C.22}$$

$$A_f = 2\pi R_f^2 \tag{C.23}$$

- Case II:

$$V_f = \frac{\pi}{3}h\left(3R_f^2 - h^2\right) \tag{C.24}$$

$$R_f = \sqrt{\frac{1}{3}\left(\frac{3V_f}{h\pi} + h^2\right)} \tag{C.25}$$

$$A_f = 2\pi R_f h \tag{C.26}$$

- Case III:

$$V_f = \int_0^c A_c\left(R_f, R_c, B, y\right) \, \mathrm{d}y + \int_c^{R_f} \pi\left(R_f^2 - y^2\right) \, \mathrm{d}y \qquad \text{(C.27)}$$

$$A_f = \int_0^c P_c\left(R_f, R_c, B, y\right) \, \mathrm{d}y + \int_c^{R_f} 2\pi\sqrt{R_f^2 - y^2} \, \mathrm{d}y \qquad \text{(C.28)}$$

- Case IV:

$$V_f = \int_0^c A_c\left(R_f, R_c, B, y\right) \, \mathrm{d}y + \int_c^{h} \pi\left(R_f^2 - y^2\right) \, \mathrm{d}y \qquad \text{(C.29)}$$

$$A_f = \int_0^c P_c\left(R_f, R_c, B, y\right) \, \mathrm{d}y + \int_c^{h} 2\pi\sqrt{R_f^2 - y^2} \, \mathrm{d}y \qquad \text{(C.30)}$$

- Case V:

$$V_f = \int_0^h A_c\left(R_f, R_c, B, y\right) \, \mathrm{d}y \qquad \text{(C.31)}$$

$$A_f = \int_0^h P_c\left(R_f, R_c, B, y\right) \, \mathrm{d}y \qquad \text{(C.32)}$$

- Case VI:

$$V_f = \frac{s\pi B^2}{4} + \int_s^h A_c\left(R_f, R_c, B, y\right) \, \mathrm{d}y \qquad \text{(C.33)}$$

$$A_f = \int_s^h P_c\left(R_f, R_c, B, y\right) \, \mathrm{d}y \qquad \text{(C.34)}$$

- Case VII:

$$V_f = \frac{s\pi B^2}{4} + \int_s^c A_c\left(R_f, R_c, B, y\right) \, \mathrm{d}y + \int_c^{h} \pi\left(R_f^2 - y^2\right) \, \mathrm{d}y \qquad \text{(C.35)}$$

$$A_f = \int_s^c P_c\left(R_f, R_c, B, y\right) \, \mathrm{d}y + \int_c^{h} 2\pi\sqrt{R_f^2 - y^2} \, \mathrm{d}y \qquad \text{(C.36)}$$

The heights of the intersections of the flame front with the cylinder walls are denoted by s and c (see Fig. C.3). The particular case of a centered ignition location can be derived by performing the limit $R_c \to 0$. Similar procedures to the one made explicit here can be found in [10, 11].

It is worth mentioning that the same kind of analysis is necessary in order to formulate a proper description of heat transfer from the working fluid to the cylinder walls. During combustion heat, transfer calculations are divided into two parts

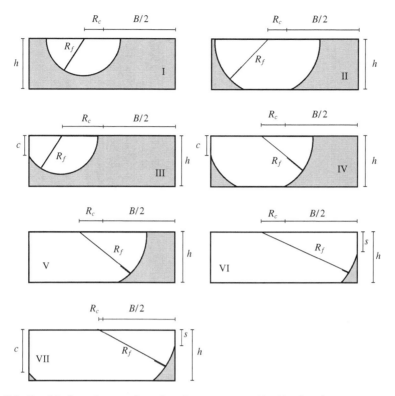

Fig. C.3 Feasible flame front configurations for a noncentered ignition location

associated either with unburned or burned gases, each one with its own temperature. Details can be found in Appendix D.

References

1. G. Beretta, M. Rashidi, J. Keck, Combust. Flame **52**, 217 (1983)
2. H. Bayraktar, O. Durgun, Energ. Sources **25**, 439 (2003)
3. J. Tagalian, J. Heywood, Combust. Flame **64**, 243 (1986)
4. C. Borgnakke, P. Puzinauskas, Y. Xiao, Spark ignition engine simulation models. Tech. rep., Department of Mechanical Engineering and Applied Mechanics. University of Michigan. Report No. UM-MEAM-86-35 (1986)
5. H. Willems, R. Sierens, in *ASME 1999 spring technical conference, ICE 32-1* (1999), pp. 83–90
6. S. Verhelst, C. Sheppard, Energ. Convers. Manage. **50**, 1326 (2009)
7. H. Bayraktar, Renew. Energ. **32**, 758 (2007)
8. N. Blizard, J. Keck, SAE Paper 740191 (1974)
9. P.L. Curto-Risso, A. Medina, A. Calvo Hernández, Appl. Therm. Eng. **31**, 803 (2011)

10. M. Modarres Razavi, A. Hosseini, M. Dehnavi, *ASME Internal-Combustion-Engine-Division Fall Technical Conference 2009, Lucerne, Switzerland*, http://www.dsy.hu/thermo/razavi/Razavi_Spark2.pdf (2010)

11. Y. Chin, R. Mattews, S. Nicholas, T. Kiehne, Combust. Sci. Technol. **86**, 1 (1992)

Appendix D
Heat Transfer Areas

For simulation purposes, heat transfer areas have to be calculated distinguishing between the cylinder wall surface wetted by the burned or the unburned gases. This is specially important during combustion because of the great difference in their temperatures. As commented in Chap. 2, for spark ignition engines the main contribution to heat transfer from the gases inside the cylinder to its wall arises from convection, so there are two contributions of the form:

$$\dot{q}_b = \frac{\dot{Q}_b}{A_b} = h_{c,g}(T_b - T_{w,g})$$

$$\dot{q}_u = \frac{\dot{Q}_u}{A_u} = h_{c,g}(T_u - T_{w,g}) \tag{D.1}$$

where it was assumed that the convective heat transfer coefficient, $h_{c,g}$, is similar for both components, unburned and burned gases. The cylinder wall areas wetted by one or the other are denoted as A_u and A_b. Most heat transfer occurs from the burned gases region. From the volume of each type of gas and considering the flame front radius as calculated in Appendix C, A_u and A_b are obtained through the equations next summarized. Two different situations are taken into account, wether the ignition is centered or not [1, 2]. For each situation, we consider exact calculations for a cylindrical combustion chamber with the same cases and notation that for the computation of the flame front area in Appendix C. Additionally, for centered ignition we mention an approximate method proposed by Ferguson [3].

D.1 Centered Ignition

D.1.1 Approximate Calculations

For centered ignition Ferguson proposes in [3], an approximate method in which it is assumed that the fraction of area contacted by burned gas is proportional to the square

A. Medina et al., *Quasi-Dimensional Simulation of Spark Ignition Engines*, DOI: 10.1007/978-1-4471-5289-7, © Springer-Verlag London 2014

root of the burned gas mass fraction, x_b. This intends to reflect the fact that the burned gas occupies a larger volume fraction of the cylinder than the unburned gas because the differences in their densities. In any case, the exponent of x_b can be considered as an adjustable parameter that can be eventually obtained from experimental measures. Equations for A_u and A_b proposed by Ferguson are:

$$A_b = \left(\frac{\pi B^2}{2} + \frac{4V}{B}\right) x_b^{1/2}$$

$$A_u = \left(\frac{\pi B^2}{2} + \frac{4V}{B}\right)\left(1 - x_b^{1/2}\right) \tag{D.2}$$

V represents the instantaneous volume of the combustion chamber and B its bore. These equations recover the correct limits in the cases $x_b \rightarrow 0$ and $x_b \rightarrow 1$ for a cylindrical volume.

D.1.2 Exact Calculations

- Case I:
$$A_b = K\pi R_f^2 \tag{D.3}$$

$$A_u = K\left(\pi \frac{B^2}{2} - \pi R_f^2\right) + 4\frac{V}{B} \tag{D.4}$$

- Case II:
$$A_b = \pi \left(2R_f^2 - h^2\right) \tag{D.5}$$

$$A_u = K\left[\pi \frac{B^2}{2} - \pi \left(2R_f^2 - h^2\right)\right] + 4\frac{V}{B} \tag{D.6}$$

- Case III:
$$A_b = K\pi \frac{B^2}{4} + \pi B\sqrt{R_f^2 - \frac{B^2}{4}} \tag{D.7}$$

$$A_u = K\left(\pi \frac{B^2}{4}\right) + 4\frac{V}{B} - \pi B\sqrt{R_f^2 - \frac{B^2}{4}} \tag{D.8}$$

- Case IV:
$$A_b = K\left[\pi \frac{B^2}{4} + \pi \left(R_f^2 - h^2\right)\right] + \pi B\sqrt{R_f^2 - \frac{B^2}{4}} \tag{D.9}$$

$$A_u = K \left[\pi \frac{B^2}{2} - \pi \frac{B^2}{4} + \pi \left(R_f^2 - h^2 \right) \right] + 4 \frac{V}{B} \tag{D.10}$$

where h the instantaneous combustion chamber height (considered as cylindrical) (see Fig. C.2), and K a phenomenological factor that arises from the fact that the cylinder head and base are not refrigerated.

D.2 Noncentered Ignition

For cylinder designs with noncentered spark plugs there exist seven possible cases, as detailed in Appendix C. It is convenient to define a function for the perimeter as:

$$P_m \left(R_f, R_c, B, y \right) = 2\alpha \left(R_f, R_c, B, y \right) \sqrt{R_f^2 - y^2} \tag{D.11}$$

where $\alpha \left(R_f, R_c, B, y \right)$ is given by Eq. (C.19).

- Case I:

$$A_b = K \pi R_f^2 \tag{D.12}$$

$$A_u = K \left(\pi \frac{B^2}{2} - \pi R_f^2 \right) + 4 \frac{V}{B} \tag{D.13}$$

- Case II:

$$A_b = \pi \left(2R_f^2 - h^2 \right) \tag{D.14}$$

$$A_u = K \left[\pi \frac{B^2}{2} - \pi \left(2R_f^2 - h^2 \right) \right] + 4 \frac{V}{B} \tag{D.15}$$

- Case III:

$$A_b = K A_c \left(R_f, R_c, B, y = 0 \right) + \int_0^c P_m \left(R_f, R_c, B, y \right) \, dy \tag{D.16}$$

$$A_u = K \left[\pi \frac{B^2}{2} - A_c \left(R_f, R_c, B, y = 0 \right) \right] + \pi B h - \int_0^c P_m \left(R_f, R_c, B, y \right) \, dy \tag{D.17}$$

- Case IV:

$$A_b = K \left[A_c \left(R_f, R_c, B, y = 0 \right) + \pi \left(R_f^2 - h^2 \right) \right] + \int_0^c P_m \left(R_f, R_c, B, y \right) \, dy \tag{D.18}$$

$$A_u = K \left[\pi \frac{B^2}{2} - A_c \left(R_f, R_c, B, y = 0 \right) - \pi \left(R_f^2 - h^2 \right) \right]$$
$$+ \pi B h - \int_0^c P_m \left(R_f, R_c, B, y \right) \, dy \qquad (D.19)$$

- Case V:

$$A_b = K \left[A_c \left(R_f, R_c, B, y = 0 \right) + A_c \left(R_f, R_c, B, y = h \right) \right]$$
$$+ \int_0^h P_m \left(R_f, R_c, B, y \right) \, dy \qquad (D.20)$$

$$A_u = K \left[\pi \frac{B^2}{2} - A_c \left(R_f, R_c, B, y = 0 \right) - A_c \left(R_f, R_c, B, y = h \right) \right]$$
$$+ \pi B h - \int_0^h P_m \left(R_f, R_c, B, y \right) \, dy \qquad (D.21)$$

- Case VI:

$$A_b = K \left[\pi \frac{B^2}{4} + A_c \left(R_f, R_c, B, y = h \right) \right] + \pi B s + \int_s^h P_m \left(R_f, R_c, B, y \right) \, dy$$
$$\qquad (D.22)$$
$$A_u = K \left[\pi \frac{B^2}{4} - A_c \left(R_f, R_c, B, y = h \right) \right] + \pi B (h - s) - \int_s^h P_m \left(R_f, R_c, B, y \right) \, dy$$
$$\qquad (D.23)$$

- Case VII:

$$A_b = K \left[\pi \frac{B^2}{4} + \pi \left(R_f^2 - h^2 \right) \right] + \pi B s + \int_s^c P_m \left(R_f, R_c, B, y \right) \, dy \quad (D.24)$$

$$A_u = K \left[\pi \frac{B^2}{4} - \pi \left(R_f^2 - h^2 \right) \right] + \pi B (h - s) - \int_s^c P_m \left(R_f, R_c, B, y \right) \, dy$$
$$\qquad (D.25)$$

All the parameters and geometrical notation are as in Appendix C.

References

1. P.L. Curto-Risso, A. Medina, A. Calvo Hernández, Appl. Therm. Eng. **31**, 803 (2011)
2. P.L. Curto-Risso, Numerical simulation and theoretical model for an irreversible Otto cycles. Ph.D. thesis, Universidad de Salamanca, Spain, http://campus.usal.es/gtfe/ (2009)
3. C.R. Ferguson, *Internal Combustion Engines, Applied Thermosciences* (Wiley, New York, 1986), Chap. 4, pp. 170–171

Appendix E
Combustion Chemistry

In this appendix, we show particular examples of how to set and solve the algebraic system of equations required to know the products after combustion, considering that the reactants include apart from fuel and air the reaction products (residual) of the previous cycle.

E.1 Hydrocarbons: Chemical Equilibrium Without Dissociation

Allow us to consider the following chemical reaction for combustion of a wet fuel with moist air and residual gases[5]:

$$\frac{(1 - y_r)}{1 + \varepsilon\phi + \tilde{\omega}} \left[\varepsilon\phi \left\{ C_a H_b O_c N_d + f\,(H_2O) \right\} + 0.21\,O_2 + 0.79\,N_2 + \tilde{\omega} H_2O \right]$$
$$+ y_r \left[y_1'\,CO_2 + y_2'\,H_2O + y_3'\,N_2 + y_4'\,O_2 + y_5'\,CO + y_6'\,H_2 \right]$$
$$\longrightarrow v_1\,CO_2 + v_2\,H_2O + v_3\,N_2 + v_4\,O_2 + v_5\,CO + v_6\,H_2 \qquad \text{(E.1)}$$

where the dry fuel is considered as an arbitrary hydrocarbon with the notation $C_a H_b O_c N_d$, carrying f moles of water per mole of dry fuel and $\varepsilon\phi$ is the amount of fuel per mole of dry air, with $\varepsilon = 0.21/(a + b/4 - c/2)$. The term $\tilde{\omega}$ is the air humidity: mole number of water per mole number of dry air.

The mole fraction of residual gases coming from the previous combustion event is denoted by y_r. The mole fraction composition of reactants in the fresh mixture is denoted by y_i', for each specie i. For each chemical specie in the right hand of the equation, the units of the multiplying coefficients are the corresponding mole number divided by the mole number of fresh mixture (subscript, u of unburned gas). For instance, units for v_1 are $mole_{CO_2}/mole_u$, units for v_4 are $mole_{O_2}/mole_u$, etc.

---

[5] In the case that traces of unburned fuel are considered in the combustion products, the fuel–air ratio has to be conveniently calculated.

These coefficients (v_i with $i = \{1, \ldots, 6\}$) are obtained by stating a set of algebraic equations arising from the conservation of the number of atoms of each specie with the following assumptions:

1. For lean mixtures, $\phi < 1$, CO and H_2 can be neglected. In this case, the set of equations is composed by four equations and, if it is considered that the residual gases and product gases have the same components as it is represented in Eq. (E.1), its solution gives all the unknowns: v_1, v_2, v_3 and v_4.

$$[C] \qquad v_1 = \frac{(1 - y_r)}{1 + \varepsilon\phi + \tilde{\omega}} [\varepsilon\phi a] + y_r y_1'$$

$$[H] \qquad 2v_2 = \frac{(1 - y_r)}{1 + \varepsilon\phi + \tilde{\omega}} \left[\varepsilon\phi (b + 2f) + 2\tilde{\omega}\right] + 2y_r y_2'$$

$$[O] \, 2v_1 + v_2 + 2v_4 = \frac{(1 - y_r)}{1 + \varepsilon\phi + \tilde{\omega}} \left[\varepsilon\phi (c + f) + 0.42 + \tilde{\omega}\right] + y_r \left(2y_1' + y_2' + 2y_4'\right)$$

$$[N] \qquad 2v_3 = \frac{(1 - y_r)}{1 + \varepsilon\phi + \tilde{\omega}} [\varepsilon\phi d + 1.58] + y_r \left(2y_3'\right)$$

$$\text{(E.2)}$$

The solution of this system of equations is:

$$v_1 = \frac{(1 - y_r)}{1 + \varepsilon\phi + \tilde{\omega}} [\varepsilon\phi a] + y_r y_1'$$

$$v_2 = \frac{(1 - y_r)}{1 + \varepsilon\phi + \tilde{\omega}} \left[\varepsilon\phi (b + 2f) + 2\tilde{\omega}\right] + 2y_r y_2'$$

$$v_3 = \frac{(1 - y_r)}{1 + \varepsilon\phi + \tilde{\omega}} [\varepsilon\phi d + 1.58] + 2y_r y_3' \qquad \text{(E.3)}$$

$$v_4 = \frac{(1 - y_r)}{1 + \varepsilon\phi + \tilde{\omega}} [0.21 (1 - \phi)] + y_r y_4'$$

$$v_5 = 0$$

$$v_6 = 0$$

2. For stoichiometric and rich mixtures, $\phi \geq 1$, O_2 can be neglected. Moreover, in this case the water gas reaction,

$$CO_2 + H_2 \rightleftharpoons CO + H_2O \qquad \text{(E.4)}$$

can be assumed to be in equilibrium. The equilibrium constant of water, $K_p(T)$, can be calculated by minimizing the Gibbs free energy, or using a polynomial fit. For instance, Heywood [1] takes:

$$\ln[K_p(T)] = 2.743 - \frac{1.761 \cdot 10^3}{T} - \frac{1.611 \cdot 10^6}{T^2} + \frac{0.2803 \cdot 10^9}{T^3} \qquad \text{(E.5)}$$

where the units of T are Kelvin.

During the simulation T, for each instant of time in the cycle, should be taken as the *adiabatic flame temperature* of the burned gases. Another simpler option is to consider K_p as approximately constant. Within this option, Heywood [1] proposes a value of 3.5 which corresponds to evaluate the equilibrium constant at 1,740 K.

The set of equations in this case includes an equation for K_p or $K_p(T)$:

$$[C] \qquad \nu_1 + \nu_5 = \frac{(1 - y_r)}{1 + \varepsilon\phi + \tilde{\omega}} \left[\varepsilon\phi a\right] + y_r \left(y_1' + y_5'\right)$$

$$[H] \qquad 2\nu_2 + 2\nu_6 = \frac{(1 - y_r)}{1 + \varepsilon\phi + \tilde{\omega}} \left[\varepsilon\phi (b + 2f) + 2\tilde{\omega}\right] + y_r \left(2y_2' + 2y_6'\right)$$

$$[O] \quad 2\nu_1 + \nu_2 + \nu_5 = \frac{(1 - y_r)}{1 + \varepsilon\phi + \tilde{\omega}} \left[\varepsilon\phi (c + f) + 0.42 + \tilde{\omega}\right] + y_r \left(2y_1' + y_2' + y_5'\right)$$

$$[N] \qquad 2\nu_3 = \frac{(1 - y_r)}{1 + \varepsilon\phi + \tilde{\omega}} \left[\varepsilon\phi d + 1.58\right] + y_r \left(2y_3'\right)$$

$$[Eq] \qquad K_p \nu_1 \nu_6 = \nu_2 \nu_5$$

$$(E.6)$$

Defining,

$$A = \frac{(1 - y_r)}{1 + \varepsilon\phi + \tilde{\omega}}$$

$$R_1 = y_r \left(y_1' + y_5'\right)$$

$$R_2 = y_r \left(2y_2' + 2y_6'\right)$$

$$R_3 = y_r \left(2y_1' + y_2' + y_5'\right)$$

$$R_4 = y_r \left(2y_3'\right) \qquad (E.7)$$

$$D_1 = A\varepsilon\phi a + R_1$$

$$D_2 = A \left[\varepsilon\phi (b + 2f) + 2\tilde{\omega}\right] + R_2$$

$$D_3 = A \left[\varepsilon\phi (c + f) + 0.42 + \tilde{\omega}\right] + R_3$$

$$D_4 = A \left[\varepsilon\phi d + 1.58\right] + R_4$$

This system of equations (E.6) admits an analytical solution that can be expressed as:

$$\nu_2 = D_3 - D_1 - \nu_1$$
$$\nu_3 = \frac{D_4}{2}$$
$$\nu_4 = 0 \qquad (E.8)$$
$$\nu_5 = D_1 - \nu_1$$
$$\nu_6 = \frac{D_2}{2} - D_3 + D_1 + \nu_1$$

which includes a quadratic equation for ν_1 that must be solved,

$$v_1^2 (K_p - 1) + v_1 \left[\left(\frac{D_2}{2} - D_3 + D_1 \right) K_p + D_3 \right] - D_1 (D_3 - D_1) = 0 \quad \text{(E.9)}$$

E.2 Hydrocarbons: Chemical Equilibrium with Dissociation

When dissociation of some species is considered, new elements appear in combustion products. The most typical and significant elements on burned gases are the six described in previous section plus four more elements: H, OH, O, and NO.

From now it is considered as the following chemical reaction for combustion of a wet fuel with moist air and residual gases:

$$\frac{(1 - y_r)}{1 + \varepsilon \phi + \tilde{\omega}} \left[\varepsilon \phi \{ C_a H_b O_c N_d + f\, (H_2O) \} + 0.21\, O_2 + 0.79\, N_2 + \tilde{\omega} H_2O \right] + y_r \left[y_1' CO_2 \right.$$
$$+ y_2' H_2O + y_3' N_2 + y_4' O_2 + y_5' CO + y_6' H_2 + y_7' H + y_8' O + y_9' OH + y_{10}' NO \right]$$
$$\longrightarrow v_1\, CO_2 + v_2\, H_2O + v_3\, N_2 + v_4\, O_2 + v_5\, CO + v_6\, H_2 + v_7\, H + v_8\, O + v_9\, OH + v_{10}\, NO$$
$$\text{(E.10)}$$

This equation do not consider fuel traces on combustion products, however, takes into account reactive elements like CO, H_2, etc. Several authors describe different methods to calculate the chemical equilibrium considering some dissociations [2–5]. This section describes a method presented by Ferguson [3] to solve chemical equilibrium for 10 species but considering a wet fuel with moist air and residual gases. All the variables used in Eq. (E.10) are defined with the same criteria that in the previous section.

The following definitions are considered:

$$A = \frac{(1 - y_r)}{1 + \varepsilon \phi + \tilde{\omega}}$$
$$R_1 = y_r \left(y_1' + y_5' \right)$$
$$R_2 = y_r \left(2y_2' + 2y_6' + y_7' + y_9' \right)$$
$$R_3 = y_r \left(2y_1' + y_2' + 2y_4' + y_5' + y_8' + y_9' + y_{10}' \right) \qquad \text{(E.11)}$$
$$R_4 = y_r \left(2y_3' + y_{10}' \right)$$
$$N = \sum_{i=1}^{10} v_i$$

and using $y_i = v_i / N$, the balance equations for each specie are:

$[C]$ $\qquad N(y_1 + y_5) = A[\varepsilon\phi a] + R_1$

$[H]$ $\qquad N(2y_2 + 2y_6 + y_7 + y_9) = A[\varepsilon\phi(b + 2f) + 2\tilde{\omega}] + R_2$

$[O]$ $N(2y_1 + y_2 + 2y_4 + y_5 + y_8 + y_9 + y_{10}) = A[\varepsilon\phi(c + f) + 0.42 + \tilde{\omega}] + R_3$

$[N]$ $\qquad N(2y_3 + y_{10}) = A[\varepsilon\phi d + 1.58] + R_4$

$$(E.12)$$

By definition it is known that,

$$\sum_{i=1}^{10} y_i - 1 = 0 \qquad (E.13)$$

In order to solve the system of equations it is necessary to introduce six equations, which are obtained from chemical equilibrium with dissociation of species. The reactions considered are the following:

$$\tfrac{1}{2}H_2 \rightleftharpoons H \qquad K_{p,1} = \frac{y_7 p^{1/2}}{y_6^{1/2}}$$

$$\tfrac{1}{2}O_2 \rightleftharpoons O \qquad K_{p,2} = \frac{y_8 p^{1/2}}{y_4^{1/2}}$$

$$\tfrac{1}{2}H_2 + \tfrac{1}{2}O_2 \rightleftharpoons OH \qquad K_{p,3} = \frac{y_9}{y_4^{1/2} y_6^{1/2}}$$

$$\tfrac{1}{2}O_2 + \tfrac{1}{2}N_2 \rightleftharpoons NO \qquad K_{p,4} = \frac{y_{10}}{y_4^{1/2} y_3^{1/2}} \qquad (E.14)$$

$$H_2 + \tfrac{1}{2}O_2 \rightleftharpoons H_2O \qquad K_{p,5} = \frac{y_2}{y_4^{1/2} y_6 p^{1/2}}$$

$$CO + \tfrac{1}{2}O_2 \rightleftharpoons CO_2 \qquad K_{p,6} = \frac{y_1}{y_4^{1/2} y_5 p^{1/2}}$$

where p represents the pressure at which the reaction occurs, in atmospheres. The equilibrium constants have the next form:

$$\log K_{p,i} = A \log\left(\frac{T}{1,000}\right) + \frac{B}{T} + C + DT + ET^2 \qquad (E.15)$$

where the required parameters are contained in Table E.1. The expression of the equilibrium constants could be rearranged to express the mole fraction of all species in terms of y_3, y_4, y_5, and y_6, the mol fractions of N_2, O_2, CO, and H_2 respectively.

Table E.1 Parameters to evaluate the equilibrium constants

	A	B	C	D	E
$K_{p,1}$	0.432168	-0.112464×10^5	0.267269×10^1	-0.745744×10^{-4}	0.242484×10^{-8}
$K_{p,2}$	0.310805	-0.129540×10^5	0.321779×10^1	-0.738336×10^{-4}	0.344645×10^{-8}
$K_{p,3}$	-0.141784	-0.213308×10^4	0.853461	0.355015×10^{-4}	-0.310227×10^{-8}
$K_{p,4}$	0.150879×10^{-1}	-0.470959×10^4	0.646096	0.272805×10^{-5}	-0.154444×10^{-8}
$K_{p,5}$	-0.752364	0.124210×10^5	-0.260286×10^1	0.259556×10^{-3}	-0.162687×10^{-7}
$K_{p,6}$	-0.415302×10^{-2}	0.148627×10^5	-0.475746×10^1	0.124699×10^{-3}	-0.900227×10^{-8}

$$y_7 = c_1 y_6^{1/2} \qquad\qquad c_1 = K_{p,1}/p^{1/2}$$
$$y_8 = c_2 y_4^{1/2} \qquad\qquad c_2 = K_{p,2}/p^{1/2}$$
$$y_9 = c_3 y_4^{1/2} y_6^{1/2} \qquad\quad c_3 = K_{p,3}$$
$$y_{10} = c_4 y_4^{1/2} y_3^{1/2} \qquad\quad c_4 = K_{p,4} \qquad\qquad \text{(E.16)}$$
$$y_2 = c_5 y_4^{1/2} y_6 \qquad\qquad c_5 = K_{p,5} p^{1/2}$$
$$y_1 = c_6 y_4^{1/2} y_5 \qquad\qquad c_6 = K_{p,6} p^{1/2}$$

The first equation of the system (E.12) is used to obtain N, then it is replaced in the remaining equations to obtain three new equations dependent on the variables previously defined.[6] These equations and Eq. (E.13) constitute a system of four equations with four unknowns:

$$2 c_5 y_4^{1/2} y_6 + 2 y_6 + c_1 y_6^{1/2} + c_3 y_4^{1/2} y_6^{1/2} - d_1 \left(c_6 y_4^{1/2} y_5 + y_5 \right) = 0 \qquad \text{(E.17)}$$

$$y_4^{1/2} \left(2 c_6 y_5 + c_5 y_6 + c_2 + c_3 y_6^{1/2} + c_4 y_3^{1/2} \right) + 2 y_4 + y_5 - d_2 \left(c_6 y_4^{1/2} y_5 + y_5 \right) = 0 \qquad \text{(E.18)}$$

$$2 y_3 + c_4 y_4^{1/2} y_3^{1/2} - d_3 \left(c_6 y_4^{1/2} y_5 + y_5 \right) = 0 \qquad \text{(E.19)}$$

with:

$$d_1 = \frac{A \left[\varepsilon \phi \left(b + 2f \right) + 2\tilde{\omega} \right] + R_2}{A \left[\varepsilon \phi a \right] + R_1}$$
$$d_2 = \frac{A \left[\varepsilon \phi \left(c + f \right) + 0.42 + \tilde{\omega} \right] + R_3}{A \left[\varepsilon \phi a \right] + R_1} \qquad\qquad \text{(E.20)}$$
$$d_3 = \frac{A \left[\varepsilon \phi d + 1.58 \right] + R_4}{A \left[\varepsilon \phi a \right] + R_1}$$

To solve this system of nonlinear equations, for instance the Newton-Raphson method could be used. The four equations can be written as:

$$f_j (y_3, y_4, y_5, y_6) = 0 \qquad j = 1, 2, 3, 4 \qquad \text{(E.21)}$$

To initialize the iteration it can be used the simple solution of the chemical equation with no dissociation (see Sect. E.1), $\left[y_3^{(1)}, y_4^{(1)}, y_5^{(1)}, y_6^{(1)} \right]$, that are reasonably close to the solution, $(y_3^*, y_4^*, y_5^*, y_6^*)$, the difference between the approximate value and the solution is:

$$\Delta y_i = y_i^* - y_i^{(1)} \qquad i = 3, 4, 5, 6 \qquad \text{(E.22)}$$

---

[6] When hydrogen fuel is used is not possible to obtain N from the first equation of the system (E.12) because there are no carbon atoms; therefore it is necessary to obtain N with other equation, i.e., the last one and rewrite the system of equations.

Expanding the left term of the Eq. (E.21) for a Taylor series and neglecting derivatives higher than first order, it can be obtained an approximate value for Δy_i,

$$f_j + \frac{\partial f_j}{\partial y_3}\Delta y_3 + \frac{\partial f_j}{\partial y_4}\Delta y_4 + \frac{\partial f_j}{\partial y_5}\Delta y_5 + \frac{\partial f_j}{\partial y_6}\Delta y_6 \approx 0 \qquad j = 1, 2, 3, 4 \qquad (E.23)$$

In matrix notation,

$$[M][\Delta y] + [f] = 0 \qquad (E.24)$$

where the terms of M are:

$$M_{11} = 1 + \frac{1}{2}\frac{c_4 y_4^{1/2}}{y_3^{1/2}}$$

$$M_{12} = \frac{c_6 y_5 + c_5 y_6 + c_2 + c_4 y_3^{1/2} + c_3 y_6^{1/2}}{2y_4^{1/2}} + 1$$

$$M_{13} = 1 + c_6 y_4^{1/2}$$

$$M_{14} = c_5 y_4^{1/2} + \frac{c_1 + c_3 y_4^{1/2}}{2y_6^{1/2}} + 1$$

$$M_{21} = 0$$

$$M_{22} = \frac{2c_5 y_6 + c_3 y_6^{1/2} - d_1 c_6 y_5}{2y_4^{1/2}}$$

$$M_{23} = -d_1\left(1 + c_6 y_4^{1/2}\right)$$

$$M_{24} = 2c_5 y_4^{1/2} + \frac{c_1 + c_3 y_4^{1/2}}{2y_6^{1/2}} + 2$$

$$M_{31} = \frac{1}{2}\frac{c_4 y_4^{1/2}}{y_3^{1/2}}$$

$$M_{32} = \frac{(2 - d_2)c_6 y_5 + c_5 y_6 + c_2 + c_4 y_3^{1/2} + c_3 y_6^{1/2}}{2y_4^{1/2}} + 2$$

$$M_{33} = (2 - d_2)c_6 y_4^{1/2} + 1 - d_2$$

$$M_{34} = c_5 y_4^{1/2} + \frac{1}{2}\frac{c_3 y_4^{1/2}}{y_6^{1/2}}$$

$$M_{41} = 2 + \frac{1}{2}\frac{c_4 y_4^{1/2}}{y_3^{1/2}}$$

$$M_{42} = \frac{c_4 y_3^{1/2} - d_3 c_6 y_5}{2y_4^{1/2}}$$

$$M_{43} = d_3 \left(c_6 y_4^{1/2} - 1 \right)$$

$$M_{44} = 0$$

To calculate the next step of the iteration, it is necessary to solve the linear system (E.24) and,

$$y_i^{(n)} = y_i^{(n-1)} + \Delta y_i \qquad i = 3, 4, 5, 6 \tag{E.25}$$

The iterations are repeated until $|\Delta y_i| < \xi \ll 1$ where ξ is a specified tolerance.

E.3 Nitrogen Oxides

The fuel combustion process in an engine requires some time to develop, but compared to the duration of a typical thermodynamic cycle, this time is considerably smaller. For this reason, as a first approximation, it can be assumed that the combustion products reach the thermodynamic equilibrium during each cycle. However, in order to quantify the emission of certain pollutants, it is necessary to consider the formation of more complex elements that may have slower oxidation times, reaching values comparable to the cycle time [4]. This is the case of the oxidation of nitrogen, which is one of the main pollutants in the case of spark ignition engines, because of its adverse effects for human health. For a detailed description of air pollution from internal combustion engines, the reader is referred to [6]. HCCI engines have the potential of reduced NO_x emissions as compared to conventional compression ignition or spark ignition engines. Several simulation works analyzed this effect [7–9].

Usually, the *nitrogen oxides* are denoted by NO_x, which represents the sum of nitrogen oxides, NO, and nitrogen dioxides, NO_2. However, the mean contribution to NO_x comes from nitrogen oxides. The temperature is the most important factor determining the rate of formation of NO. Figure E.1 shows the differences between equilibrium values of NO and the amount of NO formed at constant temperature, for four different temperatures. The end of combustion in each case is also shown. It is clear that for high temperatures, the NO formed is close to the equilibrium values, whereas for low temperatures the differences are considerable. For this reason, it is necessary to find a specific way to estimate the NO formed in the combustion process.

A procedure to calculate the ratio of formation of NO is based on the *extended Zeldovich mechanism* [10, 11] that involves three chemical reactions,

$$O + N_2 \longrightarrow NO + N$$
$$N + O_2 \longrightarrow NO + O$$
$$N + OH \longrightarrow NO + H \tag{E.26}$$

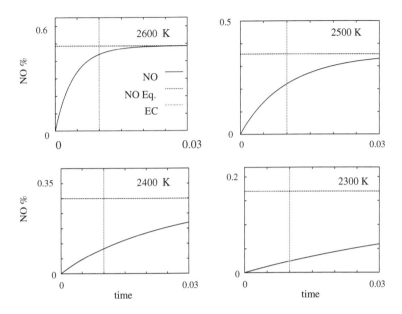

Fig. E.1 Differences between the values of the NO concentration at equilibrium (NO Eq.) and the amount of NO formed at constant temperature (NO), for an engine at 1,500 rpm, 30 bar, and four different temperatures. The end of combustion (EC) is displayed in each case as a *vertical line*

Adding the contributions of each reaction, the rate of formation of NO is,

$$\frac{d\,[NO]}{dt} = K_{f,1}\,[O]\,[N_2] + K_{f,2}\,[N]\,[O_2] + K_{f,3}\,[N]\,[OH]$$
$$- K_{r,1}\,[NO]\,[N] - K_{r,2}\,[NO]\,[O] - K_{r,3}\,[NO]\,[H] \qquad (E.27)$$

where the expressions in brackets are the mole concentration of each specie and K_i is the rate constant of each reaction, i. The subscript f refers to forward (or association) rate, and r refers to reverse (or dissociation) rate.[7]

To solve the Eq. (E.27), the concentrations of O, N_2, N, O_2, OH, and H should be estimated. The use of equilibrium values leads to fair agreement with experimental results, except for N. Since the only reactions involving N are those in Eq. (E.26) and considering that the resulting concentration is very small compared with the other species, it is possible to assume that the rate of formation of N is close to zero.

Therefore, according to Eq. (E.26), the rate of formation of N is,

---

[7] The association rate constant, K_f, and the dissociation rate constant, K_r, are related with the equilibrium constant by, $K_p = K_f/K_r$.

$$\frac{d\,[N]}{dt} = K_{f,1}\,[O]\,[N_2] - K_{f,2}\,[N]\,[O_2] - K_{f,3}\,[N]\,[OH]$$

$$- K_{r,1}\,[NO]\,[N] + K_{r,2}\,[NO]\,[O] + K_{r,3}\,[NO]\,[H] = 0 \quad (E.28)$$

Solving for [N],

$$[N] = \frac{K_{f,1}\,[O]\,[N_2] + K_{r,2}\,[NO]\,[O] + K_{r,3}\,[NO]\,[H]}{K_{r,1}\,[NO] + K_{f,2}\,[O_2] + K_{f,3}\,[OH]} \quad (E.29)$$

Substituting Eq. (E.29) into Eq. (E.27) and rearranging, we obtain the following expression for the rate of formation of NO,

$$\frac{d\,[NO]}{dt} = \frac{2C_1 \left\{ 1 - \left(\dfrac{[NO]}{[NO]_e} \right) \right\}}{1 + \left(\dfrac{[NO]}{[NO]_e} \right) C_2} \quad (E.30)$$

where $C_1 = K_{f,1}\,[O]_e\,[N_2]_e$, and:

$$C_2 = \frac{C_1}{K_{r,2}\,[NO]_e\,[O]_e\,K_{r,3}\,[NO]_e\,[H]_e} \quad (E.31)$$

The subscript e represents equilibrium conditions. To evaluate the rate constants needed in Eq. (E.30), we can consider the following expressions in $[m^3/(mol\,s)]$,

$$K_{f,1} = 7.6 \times 10^7 \exp \left(\frac{-38,000}{T} \right) \quad (E.32)$$

$$K_{r,2} = 1.5 \times 10^3\,T \exp \left(\frac{-19,500}{T} \right) \quad (E.33)$$

$$K_{r,3} = 2.0 \times 10^8 \exp \left(\frac{-23,650}{T} \right) \quad (E.34)$$

where T is evaluated in K. Equation (E.30) must be solved at each time step, together with those described in Chap. 2.

References

1. J. Heywood, *Internal Combustion Engine Fundamentals* (McGraw-Hill, New York, 1988), Chap. 4, pp. 102–105
2. C. Olikara, G. Borman, SAE Paper 750468 (1975)
3. C. Ferguson, *Internal Combustion Engines* (Wiley, New York, 1986), Chap. 3, p. 121
4. R.S. Benson, J.D. Annand, P.C. Baruah, Int. J. Mech. Sci. **17**, 97 (1975)

5. Y. Xiao, Thermodynamics properties for engine combustion simulation. Tech. rep., Department of Mechanical Engineering and Applied Mechanics, University of Michigan, Ann Arbor (1989)
6. E. Sher, *Handbook of Air Pollution from Internal Combustion Engine* (Academic Press, Elsevier, San Diego, 1998)
7. N. Komninos, D. Hountalas, Energ. Convers. Manage. **49**, 2530 (2008)
8. N. Komninos, C. Rakopoulos, Renew. Sust. Energ. Rev. **16**, 1588 (2012)
9. C. Rakopoulos, C. Michos, Energ. Convers. Manage. **49**, 2924 (2008)
10. J. Zeldovich, Acta Physiochem. URSS **21**, 577 (1946)
11. G. Lavoie, J. Heywood, J. Keck, Combust. Sci. Technol. **1**, 313 (1970)

Appendix F
Alternative Fuels

Until the beginning of the twenty-first century, petroleum reserves contributed to about 41 % of the total energy consumption of the world. Due to increasing demand for improving fuel economy and controlling air pollution, a great deal of effort has been devoted in the past few years toward designing engines with higher efficiency and lower exhaust emissions. To this end, various alternative fuels and fuel blends are being used instead of conventional fuels. In this section, we focus on describing the processes in spark ignition engines powered by alternative fuels.

It is out of the scope of this book to deeply analyze the use of different alternative fuels in spark ignition engines. Several recent reviews mainly devote to analyze the effects of considering methanol, ethanol, natural gas, hydrogen, butanol, etc., can be found in the literature [1–7]. Quasi-dimensional simulations have been widely used, together with experimental analysis, in order to deep in those effects. We refer the interested reader to Refs. [8–17] for ethanol-gasoline blends; [18–29] for natural gas; and [7, 30–41] for hydrogen.

The fundamental elements to take account in order to incorporate a fuel to the model are related with the combustion reaction, the thermodynamic properties of the working fluids, and the laminar flame speed. To solve chemical equilibrium equations, it is necessary to know the chemical composition of the fuel, that could be an unblended one or a blend of different fuels. Following the formulation presented in previous sections, the working fluid will be considered in the gaseous state; therefore the thermodynamic properties could be taken as those associated to a mixture of gases: proportional to mass rates.

F.1 Fuel Considerations for Modeling

When a pure fuel is considered, it is necessary to compile a data set in order to solve the differential equations arising from energy and mass balances. For instance, for

Table F.1 Chemical parameters for several fuels

	Gasoline (G)	Ethanol (E)	Hydrogen (H)	Propane (P)
Chemical composition:	C_8H_{18}	C_2H_5OH	H_2	C_3H_8
Stoichiometric fuel/air ratio (r_q):	0.06628	0.11137	0.0292	0.0638
Standard enthalpy of formation (J/mol K):	$-224\,010.0$	$-234\,950.0$	0.0	$-234\,950.0$
Molecular weight, M (kg/kmol):	114.2285	46.0684	2.0159	44.0956

The reference state for the standard enthalpy of formation was taken at 298.15 K

unblended isooctane,[8] ethanol, hydrogen, and propane, these parameters are given in Table F.1.

Next we analyze the case of a blend of fuels, taking as particular example a blend formed by gasoline and hydrated ethanol. If the blend is made in the liquid phase and $\tilde{y}_E$ is the percentage in volume of hydrated ethanol in the mixture and $\tilde{y}_W$ is the percentage in volume of water in ethanol, the proportion of dry fuel blended in volume, y_{df}, is:

$$y_{df} = \tilde{y}_E \left(1 - \tilde{y}_W\right) + \left(1 - \tilde{y}_E\right) \tag{F.1}$$

The amount of dry ethanol per unit of dry fuel is:

$$\widehat{y}_{dE} = \frac{\tilde{y}_E \left(1 - \tilde{y}_W\right)}{y_{df}} \tag{F.2}$$

To calculate the mole fraction of each component, it is necessary to define the number of moles in the mixture per unit of dry fuel in volume, N_m:

$$N_m = \left(1 - \widehat{y}_{dE}\right)\frac{\rho_{l,G}}{M_G} + \widehat{y}_{dE}\frac{\rho_{l,E}}{M_E} \tag{F.3}$$

where $\rho_{l,G} = 690\,\text{kg/m}^3$ and $\rho_{l,E} = 785\,\text{kg/m}^3$ represent the liquid density of gasoline and ethanol, respectively,[9] and M the molecular weight (see Table F.1). Then, the dry fuel composition in terms of the mole fractions of ethanol, y_E, and gasoline, y_G, is given by:

$$y_E = \widehat{y}_{dE}\frac{\rho_{l,E}}{M_E N_m} \tag{F.4}$$

$$y_G = \left(1 - \widehat{y}_{dE}\right)\frac{\rho_{l,G}}{M_G N_m}$$

In order to calculate the overall fuel–air ratio at stoichiometric conditions, we start from the individual combustion reactions:

---

[8] Isooctane is a reference fuel for spark ignition engines and in this section will be considered as gasoline.

[9] It is assumed that the blend is made in the liquid phase.

$$C_2H_5OH + \alpha_{s,E} \,(O_2 + 3.76N_2) \longrightarrow 2CO_2 + 3H_2O + 11.8N_2$$
$$C_8H_{18} + \alpha_{s,G} \,(O_2 + 3.76N_2) \longrightarrow 8CO_2 + 9H_2O + 47N_2 \qquad (F.5)$$

where $\alpha_{s,G} = 12.5 \,kmol_{O_2}/kmol_G$, and $\alpha_{s,E} = 3 \,kmol_{O_2}/kmol_E$ are the stoichiometric amounts of air per unit of fuel. The complete combustion reaction for a gasoline-ethanol mixture is:

$$y_G C_8H_{18} + y_E C_2H_5OH + \alpha_{s,f} \,(O_2 + 3.76N_2)$$
$$\longrightarrow [8y_G + 2y_E]\,CO_2 + [9y_G + 3y_E]\,H_2O + [47y_G + 11.8y_E]\,N_2 \qquad (F.6)$$

where $\alpha_{s,f} = 12.5y_G + 3y_E$.

The overall fuel–air mass ratio at stoichiometric conditions is the ratio between the mass of fuel and air:

$$r_q = \left(\frac{m_G + m_E}{m_a}\right)_s = \frac{y_G \, M_G + y_E \, M_E}{4.76\,\alpha_{s,f}\,M_a} \qquad (F.7)$$

It is easy to show that the overall fuel–air ratio at stoichiometric conditions of a gasoline-ethanol blend is given by:

$$r_q = \frac{y_G \alpha_{s,G} r_{q,G} + y_E \alpha_{s,E} r_{q,E}}{\alpha_{s,f}} \qquad (F.8)$$

where $r_{q,G}$ and $r_{q,E}$ are the fuel–air ratio at stoichiometric conditions of gasoline and ethanol, respectively (see Table F.1).

The standard enthalpy of formation at 298.15 K is:

$$\Delta h_f^\circ = y_E \Delta h_E^\circ + y_G \Delta h_G^\circ \qquad (F.9)$$

and the molecular weight of the blend[10]:

$$M_f = y_E M_E + y_G M_G \qquad (F.10)$$

F.2 Laminar Flame Speeds for Several Fuels

Until now, we have presented the main chemical parameters for unblended or blended fuels, taking as an example for the last a gasoline-ethanol blend. However, there is another basic parameter that should be considered in order to solve the equations for combustion: the laminar flame speed, S_L. It represents the relative velocity at which

---

[10] The water content in ethanol could be treated like an external addition of water in the chemical reaction, see Appendix E.

unburned gases move through the combustion wave in normal direction to the wave surface.

The laminar flame speed has a strong dependence on the unburned gas temperature. For example, when unburned gas temperature of an hydrocarbon is increased from 300 to 600 K, the flame speed is predicted to increase by a factor 3.64 [42]. On the other hand, experimental measurements generally show a negative dependence on pressure [43–45]. Other parameter with a strong influence is the fuel–air equivalence ratio as it is shown in Eq. (2.38). To find an adequate correlation to calculate the laminar flame speed and the corresponding exponents for pressure and temperature is probably the most difficult task in order to simulate the real behavior of an internal combustion engine using alternative fuels.

Gülder [45, 46] has written an excellent review on laminar flame speeds for alternative fuels including ethanol, isooctane ethanol, propane, etc. Using the expression (2.41) in the next form:

$$S_{L,0} = Z W \phi^\eta e^{-\xi(\phi-1.075)^2} \tag{F.11}$$

and Eq. (2.38) with the data in Table F.2, it is possible to calculate the laminar flame speed for several alternative fuels.

It is important to note that when water is present in the fuel blends, its effect should be taken into account in the correlation for S_L. Sometimes due to the lack of data, is considered as an approximation, the correlation for S_L of the corresponding anhydrous fuel.

In Sect. 3.3.3, explicit simulation results for several performance parameters in the particular case of an engine working with gasoline-ethanol blends will be shown. The comparison of simulation results with experimental ones for several concentrations of ethanol in the blend will be analyzed.

Table F.2 Laminar flame speed parameters for Eqs. (2.38) and (F.11), with $T_{ref} = 300\,\mathrm{K}$ and $p_{ref} = 100\,\mathrm{kPa}$

| | | | | | | β | |
Fuel	Z	W	η	ξ	α	$\phi \leqslant 1$	$\phi > 1$
Ethanol	1	0.465	0.25	6.34	1.75	$-0.17/\sqrt{\phi}$	$-0.17\sqrt{\phi}$
Gasoline	1	0.4658	−0.326	4.48	1.56	−0.22	
Propane	1	0.446	0.12	4.95	1.77	−0.2	
G-Ethanol	$1 + 0.07\,(\widehat{y}_{dE})^{0.35}$	0.4658	−0.326	4.48	$1.56 + 0.23\,(\widehat{y}_{dE})^{0.46}$	$\beta_E \widehat{y}_{dE} + \beta_G\,(1 - \widehat{y}_{dE})^a$	

$\widehat{y}_{dE}$ represents the volume fraction (liquid) of dry ethanol in isooctane-ethanol blends, $0 \leqslant \widehat{y}_{dE} \leqslant 0.2$

[a] Bayraktar uses a weighted sum with the amounts of ethanol and gasoline, otherwise Gülder recommends to use the hydrocarbon exponent since $\widehat{y}_{dE} < 0.2$

The use of hydrogen for powertrain systems implies one of the most important goals on CO_2 emissions reduction, due to the combustion products are only water.[11] Also, it is possible to increase the efficiency by using poor mixtures, without generating a considerable cyclic variability. In fact, the added hydrogen resulted in improved work output and a reduction in burn duration and cycle-to-cycle variation while operating under lean conditions [35]. On the other hand, the auto-ignition temperature of hydrogen is higher than gasoline, making hydrogen particularly attractive for spark ignition fuel, allowing use of high compression ratios and increasing performance [47].

The elevated cost and storage issues have done unattractive the use of this technology. However, the prices of fuel and a new environmental awareness have put on the table the need of research and development of hydrogen technology as energy carrier.

The hydrogen has an elevated energy content, its lower heating values reaches 120 MJ/kg, very high compared with 44.3 MJ/kg of isooctane. However, the density of hydrogen in the gas phase is 57 times smaller than that of isooctane. The small molecule of hydrogen presents great mass diffusion qualities, favoring the homogeneous mixture with the oxidizing and reducing the effects of heterogeneity in the combustion mixture. Besides, its wide range of flammability ($0.1 < \phi < 1.7$) allows engine power output over a wide range by varying the equivalence ratio [6]. This feature extends the limits of lean condition on spark ignition engines, since isooctane theoretical limits is around $\phi = 0.6$ much higher than $\phi = 0.1$ of hydrogen, allowing to achieve better efficiencies.

The main parameters of hydrogen to solve the differential equations are described in Table F.1. In the same way that in Sect. F.1, it is necessary to know the burning velocity. To calculate the laminar flame speed Iijima and Takeno [48] present the next correlation (at $p_{ref} = 1$ atm and $T_{ref} = 291$ K),

$$S_L = S_{L,0}\left[1 + \beta \log\left(\frac{p}{p_{ref}}\right)\right]\left(\frac{T_u}{T_{ref}}\right)^{\alpha} \tag{F.12}$$

where $S_{L,0}$, α and β are calculated by:

$$S_{L,0} = 2.98 - 1.00\,(\phi - 1.70)^2 + 0.32\,(\phi - 1.70)^3 \quad \text{(m/s)} \tag{F.13}$$

$$\alpha = 1.54 + 0.026\,(\phi - 1) \tag{F.14}$$

$$\beta = 0.43 + 0.003\,(\phi - 1) \tag{F.15}$$

Verhelst and Sierens [30] improve the correlation by including a term that considers the effect of residual gases, but for different reference ($p_{ref} = 5$ bar and $T_{ref} = 365$ K) and using an expression similar to Eq. (2.38).

---

[11] The use of hydrogen in internal combustion engines do not reduce to zero the emissions, since the high temperature, and combustion with air produce nitric oxides.

$$S_L = S_{L,0} \left(\frac{T_u}{T_{\text{ref}}}\right)^\alpha \left(\frac{p}{p_{\text{ref}}}\right)^\beta (1 - \gamma \, y_r) \tag{F.16}$$

where $S_{L,0}$, α, β and γ are calculated by:

$$S_{L,0} = -4.77\phi^3 + 8.65\phi^2 - 0.394\phi - 0.296 \tag{F.17}$$

$$\alpha = 1.232 \tag{F.18}$$

$$\beta = \begin{cases} 2.90\phi^3 - 6.69\phi^2 + 5.06\phi - 1.16 & \phi < 0.6 \\[2mm] 0.0246\phi + 0.0781 & \phi \geq 0.6 \end{cases} \tag{F.19}$$

$$\gamma = 2.715 - 0.5\phi \tag{F.20}$$

Knop et al. [49] present an alternative to calculate the laminar flame speed using Eq. (F.16), but considering a different formulation for $S_{L,0}$ and other reference ($p_{\text{ref}} = 1.013$ bar and $T_{\text{ref}} = 298$ K), so α and β are different to those in Eq. (F.18) and (F.19). For instance:

$$S_{L,0} = \begin{cases} -1.75\phi^5 + 10.81\phi^4 - 25.54\phi^3 + 26.92\phi^2 - 9.47\phi + 1.16 & 0.25 \geq \phi \geq 1.9 \\[2mm] S_{L,0}(\phi = 1.9) + \left[1.08 - S_{L,0}(\phi = 1.9)\right] \dfrac{\phi - 1.9}{5.05 - 1.9} & \phi > 1.9 \end{cases} \tag{F.21}$$

$$\beta = \begin{cases} 2.90\phi^3 - 6.69\phi^2 + 5.06\phi - 1.16 & \phi < 0.6 \\[2mm] 0.025\phi + 0.078 & \phi \geq 0.6 \end{cases} \tag{F.22}$$

$$\alpha = 1.22 \tag{F.23}$$

References

1. Y. Najjar, Open Fuels Energ. Sci. J. **2**, 1 (2009)
2. K.A. Agarwal, Prog. Energ. Combust. Sci. **33**, 233 (2007)
3. G. Timilsina, A. Shrestha, Energy **36**, 2055 (2011)
4. R. Niven, Renew. Sust. Energ. Rev. **9**, 535 (2005)
5. N. Komninos, C. Rakopoulos, Renew. Sust. Energ. Rev. **16**, 1588 (2012)
6. S. Verhelst, T. Wallner, Prog. Energ. Combust. Sci. **35**, 490 (2009)
7. G. Karim, Int. J. Hydrogen Energ. **28**, 569 (2003)
8. H. Yücescu, A. Sozen, T. Topgül, E. Arcaklioglu, Appl. Therm. Eng. **27**, 358 (2007)
9. H. Bayraktar, Renew. Energ. **30**, 1733 (2005)
10. M. Ceviz, F. Yüksel, Appl. Therm. Eng. **25**, 917 (2005)
11. M. Al-Hasan, Energ. Convers. Manage. **44**, 1547 (2003)

12. F. Yüksel, B. Yüksel, Renew. Energ. **29**, 1181 (2004)
13. H. Bayraktar, Renew. Energ. **32**, 758 (2007)
14. H. Yücescu, T. Topgül, C. Cinar, M. Okur, Appl. Therm. Eng. **26**, 2272 (2006)
15. T. Topgül, H. Yücescu, C. Cinar, A. Koca, Renew. Energ. **31**, 2534 (2006)
16. M. Celik, Appl. Thermal Eng. **28**, 396 (2008)
17. M. Koc, Y. Sekmen, T. Topgül, H. Yücescu, Renew. Energ. **34**, 2101 (2009)
18. G. Li, B. Yao, Appl. Therm. Eng. **28**, 611 (2008)
19. A. Ibrahim, S. Bari, Energ. Convers. Manage. **50**, 3129 (2009)
20. A. Sen, G. Litak, B. Yao, G. Li, Appl. Therm. Eng. **30**, 776 (2010)
21. A. Sen, J. Zheng, Z. Huang, Appl. Energ. **88**, 2324 (2011)
22. A. Sen, S. Ash, B. Huang, Z. Huang, Appl. Therm. Eng. **31**, 2247 (2011)
23. M. Lounici, K. Loubar, M. Balistrou, M. Tazerout, Appl. Therm. Eng. **31**, 319 (2011)
24. F. Ma, S. Li, J. Zhao, Z. Qi, J. Deng, N. Naeve, Y. He, S. Zhao, Int. J. Hydrogen Energ. **37**, 9892 (2012)
25. H. Zhang, X. Han, B. Yao, G. Li, Appl. Energ. **104**, 992 (2013)
26. H. Özcan, M. Söylemez, Energ. Convers. Manage. **47**, 570 (2006)
27. H. Özcan, J. Yamin, Energ. Convers. Manage. **49**, 1193 (2008)
28. S. Soylu, J. Gerpen, Energ. Convers. Manage. **45**, 467 (2004)
29. H. Bayraktar, O. Durgun, Energ. Convers. Manage. **46**, 2317 (2005)
30. S. Verhelst, R. Sierens, Int. J. Hydrogen Energ. **32**, 3545 (2007)
31. A. Sen, J. Wang, Z. Huang, Appl. Energ. **88**, 4860 (2011)
32. M. Al-Baghdadi, Turkish J. Eng. Env. Sci. **30**, 331 (2006)
33. M. Al-Baghdadi, H. Al-Janabi, Energ. Convers. Manage. **41**, 77 (2000)
34. F. Yüksel, M. Ceviz, Energy **28**, 1069 (2003)
35. T. D'Andrea, P. Henshawa, D.K. Tingb, Int. J. Hydrogen Energ. **29**, 1541 (2004)
36. C. Ji, S. Wang, Int. J. Hydrogen Energ. **34**, 7823 (2009)
37. C. Ji, S. Wang, Int. J. Hydrogen Energ. **34**, 3546 (2009)
38. S. Wang, C. Ji, Int. J. Hydrogen Energ. **37**, 1112 (2012)
39. F. Tinaut, A. Melgar, B. Giménez, M. Reyes, Int. J. Hydrogen Energ. **36**, 947 (2011)
40. F. Perini, F. Paltrinieri, E. Mattarelli, Int. J. Hydrogen Energ. **35**, 4687 (2010)
41. C. McDaid, J. Zhou, Y. Zhang, Fuel **103**, 783 (2013)
42. S.R. Turns, *An Introduction to Combustion, Concepts and Applications* (McGraw Hill, New York, 2012)
43. M. Metghalchi, J.C. Keck, Combust. Flame **38**, 143 (1980)
44. W. Xuesong, L. Qianqian, F. Jin, C. Tang, H. Zuohua, D. Ritchie, T. Guohong, X. Hongming, Fuel **95**, 234 (2012)
45. L.O. Gülder, in *SAE Paper 841000* (1984)
46. O. Gülder, in *Nineteenth Symposium on Combustion* (The Combustion Institute, Pittsburgh, 1982), pp. 275–281
47. S. Verhelst, A study of the combustion in hydrogen-fuelled internal combustion engines. Ph.D. thesis, Universiteit Gent (2005)
48. T. Iijima, T. Takeno, Combust. Flame **65**, 35 (1986)
49. V. Knop, A. Benkenida, S. Jay, O. Colin, Int. J. Hydrogen Energ. **33**, 5083 (2008)

Appendix G
Reference Geometric and Configuration Parameters for Numerical Computations

In this appendix, we collect the main parameters required to obtain numerical results for the simulations detailed Chaps. 3 and 4. Most of them were taken from the experimental setup used by Beretta et al. in [1], that we consider as a reference engine to compare with. Bayraktar et al. compile in [2] several parameters from reference engines taken from the literature (Tables G.1, G.2, G.3, G.4, G.5).

Table G.1 Reference engine operating conditions [1]

φ_0	Spark angle	330.0°
C_D	Discharge coefficient	0.6
p_{in}	Pressure at intake	0.72×10^5 Pa
T_{in}	Temperature at intake	350.0 K
p_{ex}	Pressure at exhaust	1.05×10^5 Pa
T_{ex}	Temperature at exhaust	600.0 K
$\varphi_{in,op}$	Intake valve opens	$-50°$
$\varphi_{in,cl}$	Intake valve closes	214°
$\varphi_{ex,op}$	Exhaust valve opens	490°
$\varphi_{ex,cl}$	Exhaust valve closes	750°

Table G.2 Engine geometry parameters

Cylinder-crankshaft geometry		
a	Crank radius	44.45×10^{-3} m
m_p	Piston mass	80.0×10^{-3} kg
I	Moment of inertia of the crankshaft	0.5 kg m^2
ℓ	Connecting rod length	147.0×10^{-3} m
V_0	Clearance volume	1.05×10^{-4} m^3
B	Cylinder bore	101.6×10^{-3} m
R_c	Spark location (with respect to cylinder head center)	20.0×10^{-3} m
r	Compression ratio	7.86

(continued)

A. Medina et al., *Quasi-Dimensional Simulation of Spark Ignition Engines*,
DOI: 10.1007/978-1-4471-5289-7, © Springer-Verlag London 2014

Table G.2 Continued

Valve design parameters (two valves per cylinder configuration)

w_{in}	Intake valve seat width	3.81×10^{-3} m
w_{ex}	Exhaust valve seat width	3.04×10^{-3} m
β_{in}	Intake valve seat angle	$45.0°$
β_{ex}	Exhaust valve seat angle	$45.0°$
$D_{v,in}$	Intake valve head diameter	48.3×10^{-3} m
$D_{v,ex}$	Exhaust valve head diameter	38.5×10^{-3} m
$D_{p,in}$	Intake valve inner seat diameter	39.4×10^{-3} m
$D_{p,ex}$	Exhaust valve inner seat diameter	31.41×10^{-3} m
$D_{s,in}$	Intake valve stem diameter	8.9×10^{-3} m
$D_{s,ex}$	Exhaust valve stem diameter	7.09×10^{-3} m
$L_{v,max,in}$	Intake valve maximum lift	9.3×10^{-3} m
$L_{v,max,ex}$	Exhaust valve maximum lift	9.3×10^{-3} m

Valve design parameters (four valves per cylinder configuration)

w_{in}	Intake valve seat width	1.4×10^{-3} m
w_{ex}	Exhaust valve seat width	1.4×10^{-3} m
β_{in}	Intake valve seat angle	$45.0°$
β_{ex}	Exhaust valve seat angle	$45.0°$
$D_{v,in}$	Intake valve head diameter	18.0×10^{-3} m
$D_{v,ex}$	Exhaust valve head diameter	18.0×10^{-3} m
$D_{p,in}$	Intake valve inner seat diameter	15.2×10^{-3} m
$D_{p,ex}$	Exhaust valve inner seat diameter	15.2×10^{-3} m
$D_{s,in}$	Intake valve stem diameter	4.0×10^{-3} m
$D_{s,ex}$	Exhaust valve stem diameter	4.0×10^{-3} m
$L_{v,max,in}$	Intake valve maximum lift	7.6×10^{-3} m
$L_{v,max,ex}$	Exhaust valve maximum lift	7.6×10^{-3} m

Cylinder-crankshaft geometry is as in Fig. 2.1 and valve design parameters notation as in Fig. B.1

Table G.3 Fuel characteristics

Fuel	C_8H_{18} (isooctane)
ϕ Fuel-air ratio	0.99
Fuel mass/air mass (stoichiometric)	0.0683

Table G.4 Summary of the geometrical and thermodynamical parameters used both in the theoretical results (FTT) and in the simulated ones in Sect. 3.4.4

r	Compression ratio	10
B	Bore	79.5×10^{-3} m
V_0	Clearance volume	4.96×10^{-5} m^3
T_w	Cylinder internal wall temperature	600 K
T_1	Inlet temperature (FTT)	333 K
h	Heat transfer coefficient (FTT)	1,305 W/m^2 K
m	Mass of the gas mixture inside the cylinder (FTT)	4.176×10^{-4} kg

Table G.5 Geometrical and configuration parameters considered for the simulations in Chap. 4

V_{cyl}	Cylinder volume	$8.0 \times 10^{-4}\,\text{m}^3$
a	Crank radius	$4.8 \times 10^{-2}\,\text{m}$
B	Piston bore	$9.6 \times 10^{-2}\,\text{m}$
r	Compression ratio	8
μ	Friction coefficient	16.0 kg/s
p_{in}	Intake pressure	$0.75 \times 10^5\,\text{Pa}$
T_{in}	Intake temperature	350.0 K
p_{ex}	Exhaust pressure	$1.05 \times 10^5\,\text{Pa}$
T_{ex}	Exhaust temperature	600 K

References

1. G. Beretta, M. Rashidi, J. Keck, Combust. Flame **52**, 217 (1983)
2. H. Bayraktar, O. Durgun, Energ. Sources **25**, 439 (2003)

Index

A. Medina et al., *Quasi-Dimensional Simulation of Spark Ignition Engines*,
DOI: 10.1007/978-1-4471-5289-7, © Springer-Verlag London 2014

Printed by Publishers' Graphics LLC
MLSI130828.15.16.35